YELLOWSTONE'S
REBIRTH
BY FIRE

RISING FROM THE ASHES
OF THE 1988 WILDFIRES

by Karen Wildung Reinhart

Photos by Jeff Henry and the
staff of *The Billings Gazette*

FARCOUNTRY
PRESS
HELENA, MONTANA

Acknowledgments

This book came to life because of the following people who granted me interviews and their perspectives; a special thank you goes to Roy Renkin for his endless patience with my endless questions, as well as Phil Perkins. I'm also grateful to Roger Anderson, Bob Barbee, Don Despain, Phil Farnes, Jeff Henry, Jane Lopez, Frank Markley, Leon Martindale, Kerry Murphy, Andrea Nielson, Ellen Petrick, Dave Poncin, Dan Reinhart, Ron Scharfe, Carol Shively, Conrad Smith, Lee Whittlesey, and Scott Wildung. Many more people assisted with this project through their published works, in particular Rocky Barker, Carolyn Duckworth, Mary Ann Franke, Terry McEneaney, Micah Morrison, and George Wuerthner. I am also indebted to the staffs at the Heritage and Research Center in Gardiner, Montana, and at the Jackson Hole Historical Society and Museum. I also thank Caroline Patterson, of Farcountry Press, for her editorial insight and assistance.

Photo credits:

Facing page: Despite the fact that rain had not fallen, green grass sprouted from burned ground in Elk Park, shortly after the 1988 fires. Photo by Jeff Henry, Roche Jaune Pictures.

Front cover: A firestorm at Norris on August 5, 1988. Photo courtesy of Larry Mayer, The Billings Gazette.

Title page: A rainbow arches over a thicket of young lodgepole pines between Black Sand and Biscuit Basins in 2004, sixteen years after the fires in Yellowstone. Photo by Jeff Henry, Roche Jaune Pictures.

Back cover: The North Fork Fire illuminates the night skyline south of West Yellowstone.
Photo courtesy of Larry Mayer, The Billings Gazette.

FARCOUNTRY
PRESS

ISBN 13: 978-1-56037-478-7
ISBN 10: 1-56037-478-0

For more information on our books, write Farcountry Press, P.O. Box 5630, Helena, MT 59604; call (800) 821-3874; or visit www.farcountrypress.com.

Book design and layout by Luke Duran

Library of Congress Cataloging-in-Publication Data

Reinhart, Karen Wildung.
  Yellowstone's rebirth by fire : rising from the ashes of the 1988 wildfires / by Karen Reinhart ; photos by the staff of the Billings Gazette and Jeff Henry.
    p. cm.
  ISBN-13: 978-1-56037-478-7 (softcover)
  ISBN-10: 1-56037-478-0 (softcover)
  1. Forest fires–Yellowstone National Park. 2. Forest fires–Environmental aspects–Yellowstone National Park. 3. Fire ecology–Yellowstone National Park. 4. Yellowstone National Park.
  I. Henry, Jeff. II. Billings gazette (Weekly) III. Title.
    SD421.32.Y45R45 2008
    577.2'40978752–dc22
                          2008008186

Created, produced, and designed in the United States.
Printed in Canada.

12 11 10 09 08   1 2 3 4 5 6

I dedicate this book to my children, Emma, Forrest, and Mariah,
who have grown up in Yellowstone National Park.
They continually amaze me with their capacity
to love nature and learn from her.

I also dedicate this work to everyone who fought to save
what was near and dear to their hearts in 1988, be it forests,
towns, homes, the park, or even fire itself. People love what
they love and, by their very nature, they want to keep things
the same. May they share the excitement of this time of renewal
and rebirth in Yellowstone—for this is a magical time
in the history of the world's first national park.
—Karen Reinhart

CONTENTS

Above: Stephen Albers, a sawyer on the Clover–Mist Fire, stands silhouetted against the glowing flames.
Photo courtesy of James Woodcock, *The Billings Gazette*.

Facing page: The burnt remains of lodgepole pines rise from freshly fallen snow. Photo by Jeff Henry, Roche Jaune Pictures.

Chapter One

A Meditation in Black and Green

"The fires and the recovery process have added a whole new natural dimension to all the usual attractions of Yellowstone. Instead of seeing just static natural displays, which are wonderful by themselves, people become more involved now in the natural dynamics."

—MARSHA KARLE, Yellowstone National Park Information Officer, *The New York Times*, September 3, 1989

Visitors came in droves to witness the aftermath of the 1988 Yellowstone fires. People traveling through the Roosevelt Arch at the North Entrance near Gardiner, Montana, saw blackened mountainsides before them. Fires had nearly encircled Mammoth Hot Springs, the park headquarters, just five miles up the hill, which had prompted the area's evacuation on September 10, 1988, the day before the clouds dropped snow, and the first and only time in park history that the entire park was closed to the public.

The West Entrance road was bracketed by blackened lodgepole pines, reminding people that it was at the heart of the North Fork Fire, the largest fire in 1988. Traveling north from Grand Teton National Park, visitors motored through the charred remains of forests burnt by the Huck Fire. After crossing Yellowstone's southern boundary, travelers drove nearly twenty miles through the Snake River Complex Fire that was stopped only by Yellowstone Lake at West Thumb. Travelers arriving through the Northeast Entrance encountered an escaped remnant of the Storm Creek Fire back burn that cooked right up to the edge of Cooke City, Montana.

Green, healthy, and mature forests still grew between the East Entrance and the Lake area, one of the few road segments where the views were unchanged by flame. Visitors wouldn't notice the effects of fire until they arrived at West Thumb or Canyon Village—developments that firefighters had successfully, but barely, saved. From here, their tour around the Grand Loop would be a lesson in fire: the now blackened trees stood like ebony toothpicks and, in some areas, extended as far as one could see. Visitors stopped their cars in the middle of the road and stared.

Would Yellowstone, many wondered, ever be the same?

Visitors noticed other things about Yellowstone after the fires. The sun was warming patches of ground that hadn't seen direct rays in many years. The burned and, later, toppled trees opened vistas to human eyes that hadn't been seen for probably 100 years or more. While Yellowstone may have not been easy on the eyes immediately after the fires, the views were easier to see.

Those of us who lived and worked in the Greater Yellowstone Ecosystem in 1988 were deeply affected by the monumental fires of that year. Two people lost their lives late in the fire season. Some, like outfitters

Preceding pages: Lush pre-bloom fireweed and other plants cover the scorched ground between a forest of blackened lodgepole pines, summer 1993. Photo by Jeff Henry, Roche Jaune Pictures.

Right: Visitors passing through the Roosevelt Arch at the North Entrance encounter smoky skies during the 1988 fires. Photo by Jeff Henry, Roche Jaune Pictures.

Above: During the summer of 1988, parts of Yellowstone were closed with posted signs similar to this one. Although visitors were inconvenienced with road and area closures during the fires, the park remained open to the public until September 10, 1988, when, for the first time in park history, the entire park was closed for a day. Photo by Jeff Henry, Roche Jaune Pictures.

A Meditation in Black and Green   **5**

who made their living in the area's backcountry, couldn't work. People who lived in or around the forests that had burned no longer could enjoy green vistas from their porches. Everyone suffered from unbelievable smoke, even people living at the edges of the ecosystem. Throats, gardens, yards, and wells went dry.

Visitors to Yellowstone during the fires of 1988 tell animated stories about their experiences that summer. It was an exciting, unpredictable time to be visiting the park, wrought with sudden evacuations and changes in travel plans as roads and exit gates were closed and reopened at the whim of winds and firestorms. These

Facing page: National Park Service ranger Jason Jarrett informs park visitors during a fire emergency at Nez Perce Creek in August 1988. Photo by Jeff Henry, Roche Jaune Pictures.

Below: Photographers and reporters swarm around Interior Secretary Donald Hodel at Old Faithful. Network news viewers and newspaper readers saw stories about monster wildfires, blackened forests, beleaguered tourists, suffering merchants, brave firefighters, inept public officials, flawed fire policy, and—occasionally—the fiery rebirth of nature. Photo courtesy of Larry Mayer, *The Billings Gazette*.

family stories will be passed down for years to come across the country and around the world.

Those who worked in the park over the summer saw the mosaic of black against green widen, until burned forests made up more than a third of the park. It was a difficult summer even if they realized the importance of fire to the natural ecosystem. Employees experienced 1988 as a summer of smoke-darkened skies, chaos, and bone-tired weariness, as many worked long hours with few days off. Helicopter pilots had difficulty navigating due to smoke. Helitack firefighter Jane Lopez recalled a bald eagle flying directly below their helicopter in the smoke-filled skies of 1988, and wondered if it knew where it was flying.

The fires in Yellowstone were major headlines in the national media throughout much of the long summer. Reporters dramatically discussed the burning of America's first national park, with little mention of fire ecology. They openly criticized the park's administration and fire policy, claiming that the fires

Above: An elk antler, shattered by the intense heat of a forest fire, lies on a bed of scorched Douglas fir needles near Elk Creek.
Photo by Jeff Henry, Roche Jaune Pictures.

Facing page: Young lodgepole pines burst forth among the blackened stumps of the burnt forest. Photo by Jeff Henry, Roche Jaune Pictures.

could have been tackled earlier or put out, the words themselves inciting emotional spot fires. The media held people spellbound; millions of Americans kept tabs on the burning of one of the crown jewels of the national park system.

After Yellowstone's extraordinary fires—the largest in the park's written history—park ranger naturalists (now called interpreters) had a tough job. We (I was one of them) were on the front line, taking the heat as park visitors verbally challenged park fire policy, ready to vent to anyone wearing green and gray. With equal fervor, they also wanted to know what the park was going to do with all the dead trees. Some park visitors may have been soothed if the answer had been "harvest, haul, and build houses."

Made useful, in other words. Not wasted.

It was a challenge for ecologists and naturalists to educate the public about the role of fire in an ecosystem. They had to help people see a bigger viewpoint—that the pine, spruce, and fir that were dead and down, or standing dead, had an important role in the future of the park. They would shade and protect the next generation of trees and other plants, check erosion, provide homes for birds and insects, offer cover for animals, and, perhaps most importantly, their slow decay would nourish the park's poor volcanic soil. Yellowstone's mission embraced natural processes, which included leaving trees to lie where the winds and fires set them.

Subsequent chapters will reveal the role of fire in Yellowstone, relive the stories of people who experienced that 1988 summer, and probe myths and science from the "living laboratory" of the park. Yellowstone isn't dead, devastated, or destroyed. It is reborn, rebuilt, and rejuvenated. ▣

The History of Fire in Yellowstone

"Fires of the magnitude of 1988 had happened previously and they will happen again. I lived through the event, and if I was a betting man, I'd bet I'm not going to be here when large fires in Yellowstone occur again."

—ROY RENKIN, Vegetation Management Specialist, Yellowstone National Park, 2007

Yellowstone National Park was designated the world's first national park in 1872, preserved initially for its geological wonders. Situated on top of a high volcanic plateau at the heart of the 18 million acres of the Greater Yellowstone Ecosystem, the 2.2-million-acre park is located in northwestern Wyoming and encompasses parts of Montana and Idaho. It is surrounded by Grand Teton National Park, seven national forests, three national wildlife refuges, an Indian reservation, and other public and private lands.

During the 1988 summer of fire, people experienced fire as they never had before. No one, fire experts included, remembered fire behaving as it had that summer. But that didn't mean it hadn't happened before and that it wouldn't happen again. It was an awesome and bewildering summer that compelled people to look at the element of fire in a new way.

Fire had always been present in the history of the American West. In the journals written between 1804 and 1806 by Captains Meriwether Lewis and William Clark, the explorers mentioned that the Plains Indians set small fires to improve forage for horses, to increase berry production, and to herd and hunt animals. Fires were also used to ward off enemies—fur trapper Osborn Russell, for example, noted in his journal that Blackfeet Indians set a fire in 1835 near the Madison River to drive him from hiding.

Native people did not use fire as a tool in Yellowstone to the degree that it was employed on the Great Plains, largely because Yellowstone's lodgepole pine forests do not burn readily. Sometimes called "asbestos" forests, lodgepole pine forests burn only when the conditions of drought and dry lightning are coupled with overmature trees. For example, the lodgepole pine forests that extend southeast from Norris to Lake had sprouted after a large stand-replacing fire that occurred in the 1860s. That forest fire was probably ignited by lightning rather than humans.

People began managing fires soon after Yellowstone National Park was created in 1872, when President Ulysses S. Grant signed the Organic Act into law. The park, in its early days, was mismanaged, underfunded, and overrun with poachers. Congress sent Captain Moses Harris and Company M of the First U.S. Cavalry from Fort Custer in Montana Territory to park headquarters near Mammoth Hot Springs to manage the park. When Harris and his troops arrived on August 20, 1886, one of their first jobs was to put out a forest fire burning on the mountainside near Mammoth. Harris ordered his men to grab buckets, axes, and shovels, but despite their efforts, the fire was still burning in October 1886. Snow eventually snuffed out the fire, as it did other fires in the park that year.

That was the nation's first federal firefighting effort. Harris's tactic of responding quickly to fire with a coordinated team effort is still used today. The army also introduced public campgrounds to the park in

Preceding pages: Flames leap up tree trunks near Grant Village, evidence of the tinder-dry conditions in the forested areas. Photo courtesy of Bob Zellar, *The Billings Gazette*.

Facing page: The U.S. Army leaving Mammoth Hot Springs in 1900. Yellowstone fire management began when Captain Moses Harris and Company M marched into the area in August 1886. Photo courtesy of the National Park Service.

Above: Unidentified forest fire in the Upper Geyser Basin in 1902. Photo courtesy of the National Park Service.

order to keep campfires from spreading into wildfires. Campers known as "sagebrushers" used to camp and cook wherever they pleased. Campgrounds allowed the army to contain campfires and, at the same time, to predict the location of escaped campfires that could turn into wildfires.

Up until the mid-twentieth century, virtually all fires were interpreted by people as an evil that had to be combated. People viewed fire as predatory in nature, similar to the way they viewed wolves, coyotes, and bears. They believed the animals indiscriminately killed everything in their paths, and such destructiveness was itself to be destroyed: Fires were fought; predators shot. This was the way many people experienced wild places. Nature in all her guises was to be controlled.

In Yellowstone National Park, firefighters fought all forest fires. In 1931, an 18,000-acre fire at Heart Lake was battled by 800 men. In 1940, another big fire year, 1,000 firefighters tackled fires that burned 20,700 acres. In both years, despite all of the crews' efforts, the fires were eventually extinguished by rain. After World War II, firefighting efforts improved

when helicopter pilots and smokejumpers began leading the initial response to the fires from the air.

Firefighters continued to jump on all fires until 1972, when Yellowstone administrators and fire managers embraced what was known as natural fire management. Managers drew up an experimental fire policy in Yellowstone that allowed fire to burn if it was lightning-caused and was located in one of two specially designated backcountry areas. The concept of natural management had been introduced nine years before by A. Starker Leopold, one of the primary authors of The Leopold Report, which advocated ecosystem management, in which all species and natural processes are valued equally.

People began to seriously ponder the role of predators in a healthy ecosystem. Gone were the days of bears feeding at garbage dumps or alongside roads. The reintroduction of wolves was considered. By 1973, visitors could no longer fish from the historic

Left: On guard duty, circa 1903.

Above: Soldiers performed firefighting duties throughout the park, including battling fires at the Cottage Hotel and the Double Barracks fire at Fort Yellowstone, both around 1910.

Right: Although we now know that fire is a natural part of many ecosystems, we still need to heed Smokey's message to be careful not to start a wildfire.

Below: A smokejumper parachutes in to fight a fire, 1953.

Photos courtesy of the National Park Service. Smokey Bear is used with the permission of the Forest Service, U.S. Department of Agriculture.

Top: White Lake Fire Camp A, 1953.

Above: A fire crew battles a brush fire near Mammoth Hot Springs campground, 1963.

Left: A smokejumper with firefighting equipment makes a practice jump near West Yellowstone, 1975.

Following pages: A plane drops fire retardant over the Big Teepee Fire on September 8, 1974, one of the few fires that was fought between 1972 and 1988.

Photos courtesy of the National Park Service.

Fishing Bridge. The park's mission was being reinterpreted, swinging away from human enjoyment and more toward resource protection, trying to settle on a more sustainable balance between the two. The Leopold Report not only changed the way the park interpreted its mission, it also changed the way people managed wildfire.

By 1985, managers of all public lands, including national forests, took a huge step toward acknowledging the role of fire in a healthy forest by making a commitment to allow fires to burn across park and forest boundaries. This proved challenging in 1988. The mission of the Forest Service was very different from that of the National Park Service. Whereas the National Park Service preserved natural processes in forests, including leaving dead trees prone on the forest floor, the Forest Service managed forests for "multiple uses," one of which was timber harvesting.

Between 1972 and 1988, little acreage burned in Yellowstone National Park. Lightning kindled 368

Ecologist A. Starker Leopold advocated ecosystem management, which gives equal priority to all species and natural processes. Photo by Oliver P. Pearson. Courtesy of the Museum of Vertebrate Zoology, U.C. Berkeley.

fires in the park. More than 60 percent of these fires were allowed to burn; only 10 percent of which burned more than one acre. The most acreage burned was 21,000 acres—1 percent of the park—in 1981. Fires in Yellowstone National Park, at least large fires, didn't burn readily.

All that changed in 1988. At the beginning of the 1988 fire season, natural fire management policy prevailed. Lightning-caused fires were allowed to burn in places where they did not threaten human life, property, historical or cultural sites, and specific natural features, or where they did not threaten and endanger wildlife. Human-caused fires were to be suppressed using a "light on the land" approach, which meant using hand tools and firefighters to fight fires, but not bulldozers. If needed, prescribed burns could be used to reduce hazardous fuels.

As the fire season progressed, attitudes changed and all fires were suppressed. The perfect conditions for a large fire had come together: drought, dry lightning, and over-mature forests. ⊠

August 20: **A Dark Day in Fire History**

August 20, 1886: Captain Moses Harris and Company M of the First U.S. Cavalry conducted the first organized firefighting effort in the American West, when they attempted to put out a fire near park headquarters at Mammoth Hot Springs in Yellowstone National Park. Harris's firefighting tactics of an organized response to a fire are still used today.

August 20, 1910: The fateful beginning of the "Big Blowup." To escape a raging firestorm burning near the border of Idaho and Montana, ranger and fire crew leader Ed Pulaski led forty-five firefighters to the War Eagle Mine in the Coeur d'Alene National Forest near Wallace, Idaho. Early the next morning, after his crew thought he was dead, he emerged from the mine shaft blind, but he and thirty-nine other survivors walked to safety. During the two-day blowup, eighty-five people perished, a million acres of forest burned in Idaho and Montana, and towns were wiped off the map. Pulaski later fashioned a tool that combined axe and hoe, a firefighting tool that firefighters still use today.

August 20, 1988: This is the day known to Yellowstone firefighters, residents, and visitors as "Black Saturday." Forecasters predicted a red-flag day of hot temperatures, low humidity, and gale-force winds. Nature didn't disappoint. Winds gusted up to 80 mph, whipping the major fires in the Greater Yellowstone Ecosystem into an unprecedented frenzy, making firefighting nearly impossible by land or air. Trees toppled and fires grew so fierce they created their own unpredictable winds. By midnight, 165,000 additional acres had burned within Yellowstone National Park.

Lightning, Wind, and Firefighters

THREE

"I hated the way the media used 'let it burn' to describe Yellowstone's fire policy. We were getting our teeth kicked in. It's not like we had our feet up on the desk!"

—Phil Perkins, Fire Information Officer, 2007

The Greater Yellowstone Ecosystem fires of 1988 resulted from a convergence of climatic and weather conditions. The unusually low snowfall during the previous two winters set the stage for the fires. Ample snowpack is critical to maintaining adequate underground moisture and to keep rivers and streams flowing, although annual spring and summer rains also contribute to the overall moisture. The snowpack at the headwaters of the Yellowstone River on April 1, 1988, was 70 percent of the long-time average. One month later, it was 69 percent of the average. In June, when precipitation would have mostly been in the form of rain, only 40 percent of the average precipitation fell. Streams and rivers slowed. Fallen trees in the forest became drier than kiln-dried lumber or dresser drawers at home. Plants dried out. Grasses crackled.

Fire behaviorists—experts who study fire and

Preceding pages: Flames consume one of nineteen buildings destroyed when the North Fork Fire overran the Old Faithful Complex. Most of the buildings burned were small cabins.
Photo courtesy of Robert Ekey, *The Billings Gazette.*

Right: Monster flames explode over an unburned forest.
Photo courtesy of Robert Ekey, *The Billings Gazette.*

Below: Rick Hutchinson, a National Park Service geologist, checks the wind speed at Gibbon Meadows during the Wolf Lake Fire.
Photo by Jeff Henry, Roche Jaune Pictures.

predict its advance—were nicknamed "fire gods." They had only to look at the weather history to predict that rain would come in July. Over the previous eleven years in Yellowstone, precipitation in July had averaged a whopping 183 percent of normal. The five years preceding 1988 had been especially wet. As late as July 11, 1988, the National Weather Service predicted normal July rainfall for Yellowstone National Park. The rains would come.

The Heat of Summer

The summer was hot and dry. At least six different cold fronts whipped across the Yellowstone landscape, bringing lightning, but not rain. Instead, ferocious winds were their dance partners, bowling over stands of young trees, sucking moisture out of everything,

Below: Firefighters gather at the Gardiner High School, just outside the park's North Entrance. Photo by Jeff Henry, Roche Jaune Pictures.

Facing page: Firefighters line up to march, pulaskis in hand, to halt the spread of the Clover–Mist Fire. Photo courtesy of Bob Zellar, The Billings Gazette.

fanning blazes beyond anyone's wildest imaginings.

In the summer of 1988, eight major blazes burned within Yellowstone National Park. The first major fire, the Storm Creek Fire, was ignited by lightning on June 14, in the Custer National Forest beyond the park's northeast corner. By the end of July, lightning had started three more fires inside park boundaries—the Snake River Complex (Falls, Shoshone, and Red), Fan, and Clover–Mist fires. Two blazes had begun in nearby national forests: the Mink Fire in the Bridger–Teton National Forest southeast of the park and the North Fork Fire in the Targhee National Forest near the park's southwest boundary; the latter was caused by a discarded cigarette. Two more began in August 1988— the Hellroaring Fire was started by a spark from a horseshoe near an outfitter's camp in the Gallatin National Forest north of the park; the Huck Fire was torched by a tree that fell on a power line on the John D. Rockefeller, Jr. Memorial Parkway, south of Yellowstone.

Fire experts and firefighters were surprised more than once during the summer of fire. Firebrands, or

Above: Incident Commander Dave Poncin reviews a map at the Madison Fire Camp with National Park Service Director William Penn Mott.
Photo by Jeff Henry, Roche Jaune Pictures.

Right: Signs demarcate the Fountain Flats helibase.
Photo by Jeff Henry, Roche Jaune Pictures.

Facing page: Yellowstone Superintendent Bob Barbee at Old Faithful, during a review of firefighting equipment, July 27, 1988. Behind Barbee is park spokesperson Joan Anzelmo. Photo by Jeff Henry, Roche Jaune Pictures.

balls of fire, shot as far as a mile and a half ahead of the fire front on strong, horizontal winds, starting new fires and rendering human-made firebreaks useless. Even natural firebreaks, such as the half-mile-wide Grand Canyon of the Yellowstone, didn't hold. And, unlike other fires, where the flames would quiet at night, Yellowstone skies harbored an eerie reddish glow at night, as high temperatures and low humidity kept fires from "lying down." Firefighters can usually make progress against fire during evening hours, but this did not hold true in 1988.

Crews in Yellowstone began fighting the fires in mid-July. A week later, a Unified Area Command was established, employing fire experts from the National Park Service and the Forest Service. Firefighters, bulldozers, fixed-wing aircraft, and helicopters attacked the fires from the ground and above. By mid-August, Yellowstone was considered to be under siege: the military was called in to give firefighters a reprieve. The U.S. Army, Navy, Air Force, Marines, and the Wyoming National Guard battled the blaze. Over the course of the summer, 25,000 people were involved in the firefighting effort—9,600 at one time—including 11,700 military personnel, at a cost of more than $120 million. No people died in the park due to the fires, but outside the park two firefighters lost their lives battling the blazes.

The heroic effort of humans—firefighters digging fire lines, pilots using various aircraft to douse flames with water and retardant, incident commanders

safely leading crews with their expertise, and the park superintendent making the big decisions that guided the effort—successfully protected human life and most structures. But money, muscles, and machines could not stop the flames. People were humbled by the power and the awful necessity of fire.

It was a summer of controversy, fury, and sensational media coverage. People didn't understand the role of fire in the ecology of Yellowstone and the surrounding forests, and viewed the park's burning as sheer devastation rather than the beginning of the park's regeneration. People were angry about what they viewed as the horrors of the Yellowstone fires, distraught that the park they loved would never be the same. Blame was widespread: The public held the park superintendent, the director of the National Park Service, and other public land managers responsible for what many viewed as a national catastrophe. Politicians wanted them fired.

The summer of 1988 was the driest in the 112-year recorded history of the park. Only 36 percent of the annual average rainfall fell during June, July, and August. After the fires were finally snuffed out by a thick carpet of snow in mid-November, 36 percent

Facing page: Sun shines through heavy smoke behind a parked helicopter at a helibase in the Lamar River Valley. Photo by Jeff Henry, Roche Jaune Pictures.

Below: The U.S. Army, Navy, Air Force, Marines, and the Wyoming National Guard help battle the Yellowstone fires.
Photo courtesy of Larry Mayer, *The Billings Gazette*.

of the Yellowstone landscape—793,880 acres—had burned with varying intensity.

In the Greater Yellowstone Ecosystem, about 1.4 million acres of land were scorched. Three of the eight major fires had been caused by humans and accounted for approximately half of the total acreage burned in the ecosystem. The five fires that began outside of Yellowstone's boundaries accounted for more than 60 percent of the acreage that burned within the park. Sixty-seven structures were destroyed, including eighteen backcountry cabins; property damage totaled more than $3 million. There were measureable casualties among large mammals as well: 345 elk, 36 deer, 12 moose, 9 bison, and 6 black bears.

Yellowstone Park had been forged anew by fire.

The Joy of Snow

The heat and intensity of the summer fires began to give way to cooler temperatures when the season's first snow flurries began to fall on September 11. A collective sigh of relief might have been heard from the firefighters in Yellowstone and from people across the nation. Fires began to hiss, shushed by the gift of rain and snow that was ushered in by the same dervish winds that had hurled the North Fork Fire toward park headquarters and forced the park's evacuation and the first complete closure in the park's history. Miraculously, the windstorm also delivered precipitation, a blessing that hadn't been bestowed on Yellowstone National Park virtually all summer.

The snow was the beginning of the end. Though the saga of the 1988 fires wouldn't be over until the last weary firefighter went home after "mopping up" the remnants of fire, or until after the last review or report was written, the epic fire season was coming to a blessed close. But the fires of 1988 will never be completely forgotten. They will be kept alive by those who experienced them—firefighter, ecosystem resident, visitor, and television viewer—because the fires of 1988 changed the way we look at fire. ☒

Above: Research biologist Don Despain takes photographs of fire behavior. Photo by Jeff Henry, Roche Jaune Pictures.

Facing page: A spike bull elk feeds on vegetation left unburned by the North Fork Fire during the first snowfall at Gibbon Meadows on September 11, 1988. Photo by Jeff Henry, Roche Jaune Pictures.

Facing page: Some deer, elk, and moose were trapped in quickly moving firestorms and died. Photo courtesy of James Woodcock, *The Billings Gazette*.

Above: A military C-130 tanker drops retardant on the North Fork Fire. Photo courtesy of Larry Mayer, *The Billings Gazette*.

1988 Fire Statistics

Nature's Tally

793,880	Acres burned (36 percent) of Yellowstone National Park
1,400,000	Acres burned in the Greater Yellowstone Ecosystem
30,000	Acres of harvestable timber burned in nearby national forests
$12 million	Value of 30,000 acres of timber
$21 million	Cost of harvesting 30,000 acres of timber

Property

67	Mobile homes and private and government-owned cabins destroyed (12 additional structures badly damaged)
300	Utility poles and more than 10 miles of power lines damaged or destroyed

Animals

408	Large mammals perished by fire (including 345 elk, 36 deer, 12 moose, 9 bison, 6 black bears) and numerous small animals

Money Spent

$120,000,000	Total firefighting expenditures ($33 million on services and non-government help, mostly in communities near the park)

On the Ground

25,000	People (including nearly 11,700 military personnel) fought fires; 9,600 firefighters at one time
2	Deaths (a pilot and a firefighter died outside the park)
1	Accident (a helicopter crashed in the park, but there was no serious injury; other injuries and illnesses were minor)
11	Fire situations overran or threatened firefighters (forcing deployment of nearly 4 dozen emergency fire shelters)
665	Miles of fire line dug
137	Miles of fire line bulldozed
100+	Fire engines used (some from as far away as Colorado and South Dakota)

In the Air

100+	Aircraft employed (logging 18,000 flight hours)
77	Helicopters employed (carrying more than 10 million gallons of water to put out fires)
1,400,000	Gallons of fire retardant dropped by planes

Yellowstone National Park

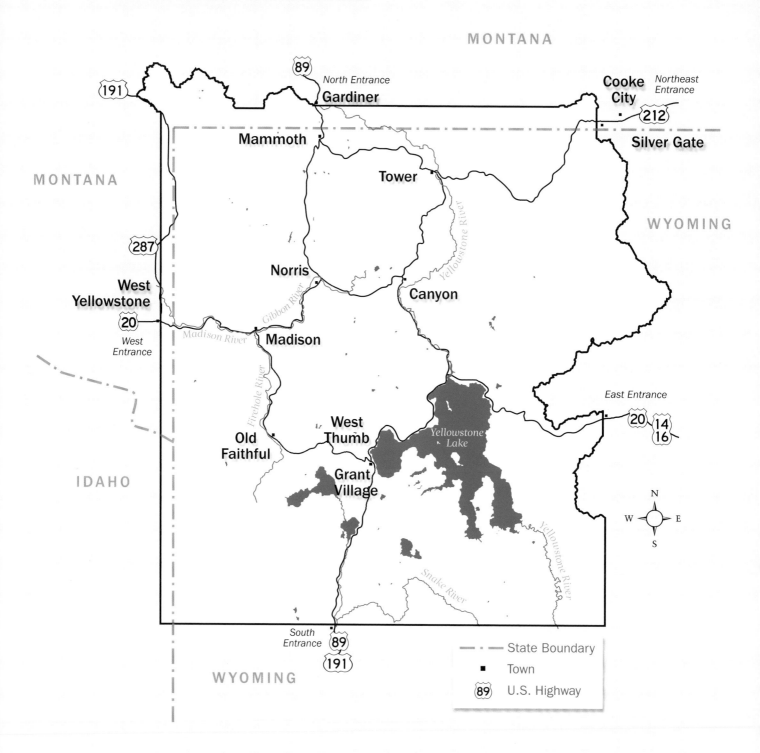

Above: Yellowstone National Park is at the heart of the 18-million-acre Greater Yellowstone Ecosystem. The 2.2-million-acre park is located in northwest Wyoming and encompasses parts of Montana and Idaho. Map illustration by Luke Duran, Element L Design.

Final Perimeters of 1988 Yellowstone Ecosystem Fires

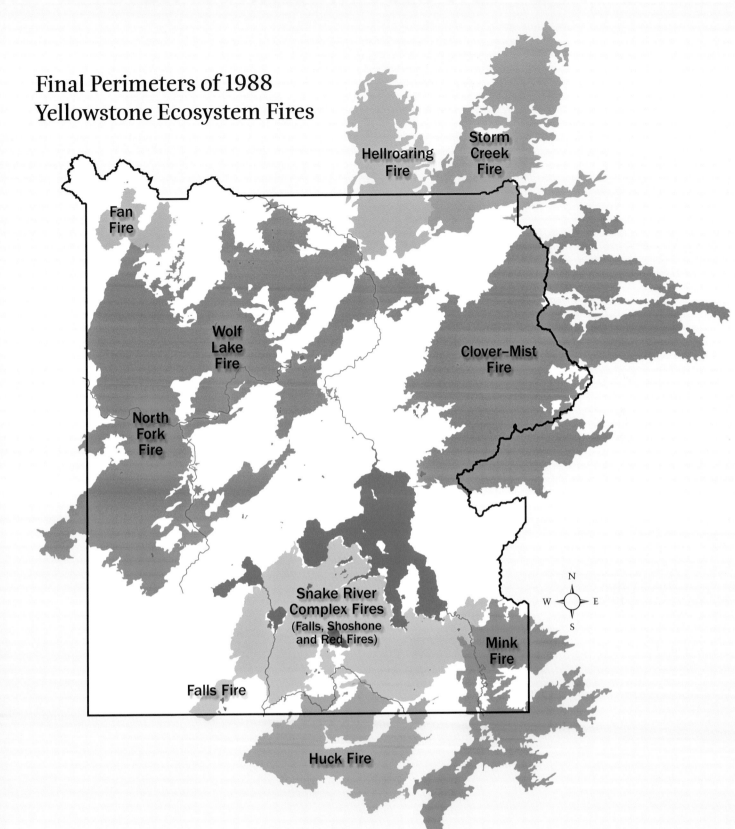

Hellroaring Fire

Storm Creek Fire

Fan Fire

Wolf Lake Fire

Clover–Mist Fire

North Fork Fire

Snake River Complex Fires (Falls, Shoshone and Red Fires)

Mink Fire

Falls Fire

Huck Fire

N
W E
S

Above: Fire knows no boundaries. More than half of the acreage that burned inside the park was ignited by humans or began beyond Yellowstone's borders. Nearly 800,000 acres burned inside park boundaries—approximately 36 percent of the park—and 1.4 million acres burned within the ecosystem.

Map illustration by Luke Duran, Element L Design.

Chapter Four

Timeline of a Hot Summer

FOUR

"It's ecological arrogance to say that we should go out here and manage this—crop the herds, pre-burn the forests, manipulate Mother Nature, with some idea that everything can be kept just as it was. Impossible! Everything keeps changing, the climate, the animal cycles. The pine bark beetle, beaver, aspen—everything comes and goes!"

—DON G. DESPAIN, Yellowstone Research Biologist and Plant Ecologist, *The New York Times*, December 11, 1988

For the people in Yellowstone, the summer of 1988 was a blur. Days stretched into endless weeks and seemed to merge together like the fires themselves. Day after day, firefighters, volunteers, and park staff faced crises of one degree or another. There was little rest, fresh air, or certainty to help with the physical and mental challenges. The following timeline highlights the biggest days of the 1988 fire season in Yellowstone, but during the heat and the smoke of that long summer, no one knew which days these would be.

June 1988

The Storm Creek Fire was started by lightning on June 14, in the Custer National Forest northeast of Yellowstone National Park. It wasn't the first fire of the summer—there had been other lightning strikes in

Preceding pages: Fire trucks gather and a herd of elk grazes as smoke clouds loom over Mammoth Hot Springs, park headquarters, in September 1988. Photo courtesy of Larry Mayer, *The Billings Gazette*.

Facing page: Firefighter Chris Glenn sharpens tools on a grinder at the Old Faithful Fire Camp. Photo by Jeff Henry, Roche Jaune Pictures.

May, but the rain had promptly doused them. On June 23, at the opposite end of the Greater Yellowstone Ecosystem, Mother Nature spoke again when lightning ignited a fire near Shoshone Lake, inside park boundaries, and started the Shoshone Fire. Two days later, another lightning strike ignited the Fan Fire, along U.S. 191 in the park's northwest corner. As they had done with backcountry fires for the previous twenty years, park fire managers let these natural fires burn as long as they did not threaten people or property.

Though precipitation in June had been lacking, weather experts predicted that rains would come in July.

July 1–14, 1988

The first day of July, the Red Fire was born from a lightning strike near Lewis Lake in the park's southern region. Eight days later, on July 9, the Mist Fire sprang up just inside the park's eastern border after a dry thunderstorm passed over. Since June, nature seemed to be playing a game of ping-pong with fire—batting fireballs back and forth over the park's high plateau, sparking major fires on opposite ends of the park.

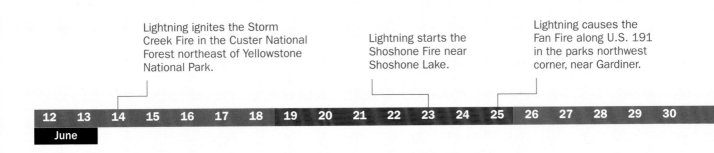

Lightning ignites the Storm Creek Fire in the Custer National Forest northeast of Yellowstone National Park.

Lightning starts the Shoshone Fire near Shoshone Lake.

Lightning causes the Fan Fire along U.S. 191 in the parks northwest corner, near Gardiner.

| 12 | 13 | 14 | 15 | 16 | 17 | 18 | 19 | 20 | 21 | 22 | 23 | 24 | 25 | 26 | 27 | 28 | 29 | 30 |

June

Two lightning-caused fires that would become major fires were ignited on July 11: the Clover Fire near the eastern border of Yellowstone and the Mink Fire south of the park in the Bridger–Teton National Forest. On July 12, yet another fire, the Falls Fire, started just east of the park's south entrance.

At this point, Yellowstone's fire information officers were struggling to find people to monitor the fires. Even though fires were initially allowed to burn, the park still needed people to monitor the fires' behaviors and to assess fuel moisture levels in the forest. The Forest Service couldn't help the Park Service by lending their people or money until the fires were

declared "wildfires" and firefighting was underway. Until then, Yellowstone had to bring in people from as far away as Alaska to observe the fires.

On July 14, the 300-acre Clover Fire consumed more than 4,000 acres of forest in a few hours, inciting the first human drama of the fire season. A three-person crew had been clearing the trail before Vice President George W. Bush's scheduled horseback ride into the park's backcountry from the Shoshone National Forest east of the park. Earlier that day, Superintendent Bob Barbee had called the supervisor of the Shoshone National Forest, telling him to cancel the future president's trip to the park. Bush became the

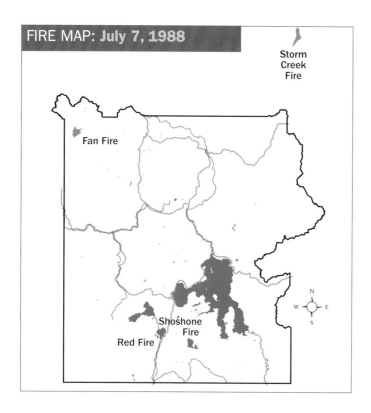

FIRE MAP: July 7, 1988

Storm Creek Fire

Fan Fire

Shoshone Fire

Red Fire

N
W E
S

Above: Amy Recker, Bonnie Gafney, and an unidentified firefighter take a break from fireproofing the Nez Perce Patrol Cabin in the Lower Geyser Basin. Photo by Jeff Henry, Roche Jaune Pictures.

Facing page: Park employee Sheila Lawson gazes at a giant cloud thrown up by the North Fork Fire on August 11, 1988. Photo by Jeff Henry, Roche Jaune Pictures.

This Mist Fire is started after a thunderstorm on the eastern edge of Yellowstone.

The Clover Fire is ignited by lightning near the eastern border of the park.

The Clover Fire expands from 300 acres to more than 4,000 acres in just a few hours.

July

| 1 | 2 | 3 | 4 | 5 | 6 | 7 | 8 | 9 | 10 | 11 | 12 | 13 | 14 |

Lightning ignites the Red Fire near Lewis Lake.

Lightning ignites the Mink Fire in the Bridger–Teton National Forest south of the park.

first park visitor who was turned back that summer.

When Chief Ranger Dan Sholly realized the inferno threatened the Calfee Creek Cabin in the Lamar River drainage, he changed the trail crew's mission to cabin protection and transported them the three miles to Calfee Creek. Sholly and two crew members huddled inside two emergency fire shelters during the firestorm that developed.

The next day, park officials issued the park's first official fire map.

July 15–23, 1988

On July 17, National Park Service firefighters began fighting fire in Yellowstone National Park. Even though it hadn't yet been designated a full-blown wildfire, crews attacked the south flank of the Falls Fire to keep it from burning into the Targhee National Forest. On July 21, after 17,000 acres had burned in the park, officials realized they weren't facing an ordinary fire season. They changed the fire management strategy from a "let it burn" fire policy to full suppression of all fires regardless of cause, and the Clover, Mist, and Snake River Complex fires (the combined Red, Shoshone, and Falls fires) were officially declared wildfires. The firefighting had begun.

On July 22, a man who was cutting firewood in the Targhee National Forest just outside the park's West Entrance carelessly discarded a cigarette and touched off a fire that engulfed nearly 500 acres in six hours. Forest Service firefighters aggressively fought the fire

Facing page: A fireball explodes skyward after a dry hillside erupts in flame east of Cooke City, September 10, 1988.
Photo courtesy of James Woodcock, *The Billings Gazette*.

from the beginning, but it quickly spread into the park. On the same day, the Clover and Mist fires merged to become the Clover–Mist Fire on the park's east side.

On July 23, 4,000 people were forced to evacuate Grant Village because of the approaching Shoshone Fire. Lorraine Mintzmyer, Rocky Mountain regional director, called in a Unified Area Command, a two-person team of National Park Service and Forest Service fire experts that would coordinate and oversee all firefighting efforts in Yellowstone. From the small town of West Yellowstone, the fire management team set up a command post where they made decisions about the disbursement of firefighters, equipment, and supplies, and that served as a communications hub for agencies, the media, and the public.

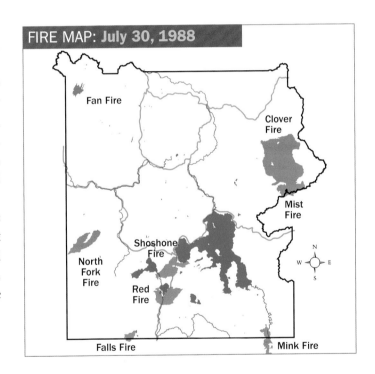

FIRE MAP: July 30, 1988

Fan Fire

Clover Fire

Mist Fire

Shoshone Fire

North Fork Fire

Red Fire

Falls Fire

Mink Fire

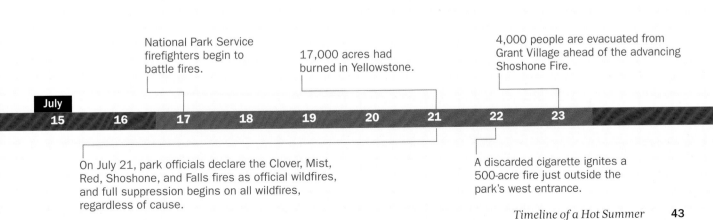

National Park Service firefighters begin to battle fires.

17,000 acres had burned in Yellowstone.

4,000 people are evacuated from Grant Village ahead of the advancing Shoshone Fire.

July

15 16 17 18 19 20 21 22 23

On July 21, park officials declare the Clover, Mist, Red, Shoshone, and Falls fires as official wildfires, and full suppression begins on all wildfires, regardless of cause.

A discarded cigarette ignites a 500-acre fire just outside the park's west entrance.

Top right: Firefighters reduce potential fuel sources by removing fallen logs near Canyon.
Photo by Jeff Henry, Roche Jaune Pictures.

Bottom left: Firefighters Mark Haroldson and Mark McCutcheon on top of Canyon Lodge work as "spotters," watching for the fire's approach.
Photo by Jeff Henry, Roche Jaune Pictures.

Bottom right: A fire crew reduces fuels to create a safe perimeter around a patrol cabin.
Photo by Jeff Henry, Roche Jaune Pictures.

July 24–31, 1988

On July 25, the story of the Yellowstone fires became national news. Newspaper headlines across the country blasted the "let it burn" fire policy, although the firefighters had been aggressively fighting all fires in Yellowstone for four days. After firefighters cut down trees and cleared debris from around structures in Grant Village, the Shoshone fire detoured around the development on July 26. The fire turned north toward West Thumb, seeking fuels and propelled by high winds.

On July 27, Interior Secretary Donald Hodel held a press conference at Old Faithful Inn and expressed support for the park's fire management policy, applauding the benefits of fire in the ecosystem and hoping to calm public furor. At this point, fires had burned nearly 89,000 acres. On the last day of the month, the Shoshone fire made a run for West Thumb. To protect the area, firefighters removed fuels by cutting down trees next to buildings and burning down isolated stands of trees. All of West Thumb's structures were saved.

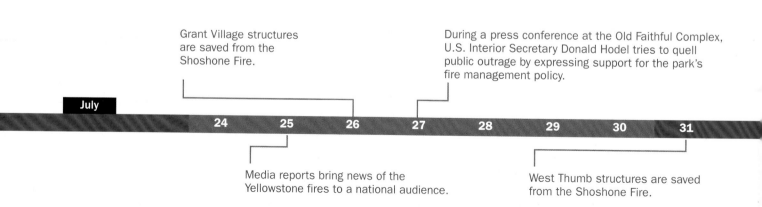

Grant Village structures are saved from the Shoshone Fire.

During a press conference at the Old Faithful Complex, U.S. Interior Secretary Donald Hodel tries to quell public outrage by expressing support for the park's fire management policy.

July | 24 | 25 | 26 | 27 | 28 | 29 | 30 | 31

Media reports bring news of the Yellowstone fires to a national audience.

West Thumb structures are saved from the Shoshone Fire.

August 1–7, 1988

On August 2, firefighters positioned bulldozers on the northwestern border of the park as the Fan Fire threatened lands owned by the Church Universal and Triumphant, an eclectic religious sect that was headquartered on a ranch north of Yellowstone. As fire approached, 250 church members and sect leader Elizabeth Clare Prophet chanted "Reverse the tide. Roll them back. Set all free." The following night, the fire activity subsided and didn't cross the park boundary; the firefighters and church members all claimed credit. By the end of the week, nearly 1,500 firefighting personnel were working to keep the Fan Fire inside park boundaries.

August 8–14, 1988

In the park's southern end, near Shoshone Lake, the Shoshone and Red fires merged into a single confla-gration on August 8. By August 14, the North Fork Fire was 53,000 acres strong; the Red–Shoshone was mapped at 55,000 acres. The largest fire at this point in time was the Clover–Mist at 95,000 acres.

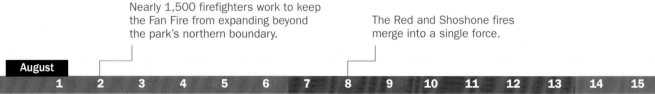

Nearly 1,500 firefighters work to keep the Fan Fire from expanding beyond the park's northern boundary.

The Red and Shoshone fires merge into a single force.

August

| 1 | 2 | 3 | 4 | 5 | 6 | 7 | 8 | 9 | 10 | 11 | 12 | 13 | 14 | 15 |

By August 14, the North Fork Fire had burned 53,000 acres, the Red–Shoshone Fire had expanded to 55,000 acres, and the Clover–Mist Fire had burned 95,000 acres.

Above: TW Services employee
Sue Harn shields her face from
flying ash and embers as the
firestorm hits Old Faithful.
Photo by Judy Tell, *The Billings Gazette*.

Right: A campground sign at the
Northeast Entrance warns visitors
that the majority of the park
is closed to camping.
Photo by Jeff Henry, Roche Jaune Pictures.

Facing page: As they stand against
a flame-colored sky, Yellowstone
National Park officials—(from left)
Denny Sutherland, John Lounsbury,
and Roger Andrascik—discuss the
intensive fire activity around them on
Black Saturday, August 20, 1988.
Photo by Larry Mayer, *The Billings Gazette*.

August 15–21, 1988

On August 15, the Hellroaring Fire began when a spark from a horseshoe ignited brush near an outfitter's camp in the Gallatin National Forest north of Yellowstone. At the same time, the Clover–Mist Fire in the northeast corner of the park pressed toward Cooke City and Silver Gate. On August 19, the Boise Interagency Fire Center called the U.S. Department of Defense and requested emergency firefighters because their firefighter reserve was empty. Also by this date, the fires in Yellowstone had spread to over 300,000 acres—not a good time to run short on firefighters. Yellowstone wasn't the only hot spot in the West, and it was competing for resources with other fires that were closer to large population centers.

August 20 was "Black Saturday." It was a good day for fire but a bad day for firefighters—virtually all firefighting efforts were rendered ineffective because of the extreme and unpredictable winds. All eight fires made major runs, propelled by winds that gusted up to 80 mph and hurled fireballs nearly one mile ahead of the flames. Some firefighters fled from the flames that day; others could only stand by in a safe zone, watching as fire rumbled, then roared through the forest on its own wind, sucking oxygen

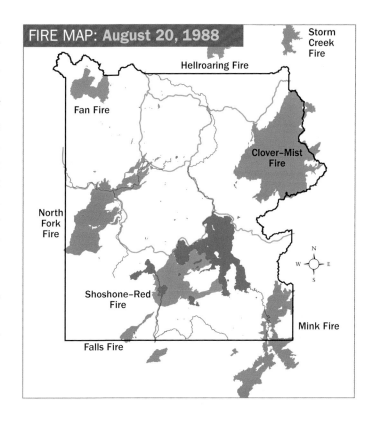

FIRE MAP: August 20, 1988

Storm Creek Fire
Hellroaring Fire
Fan Fire
Clover–Mist Fire
North Fork Fire
Shoshone–Red Fire
Mink Fire
Falls Fire

Left: Powered by 80 mph wind gusts, a wall of flame from the North Fork Fire rages toward the Norris Geyser Basin on Black Saturday, August, 20, 1988. Photo courtesy of Larry Mayer, *The Billings Gazette*.

Following pages: A Forest Service crew drives through Cascades Meadow chasing spot fires as the Wolf Lake Fire approaches Canyon Village. Photo courtesy of Bob Zellar, *The Billings Gazette*.

A giant Vertol helicopter swoops overhead carrying a 1,000-gallon bucket. Photo by Jeff Henry, Roche Jaune Pictures.

from along the ground to sustain itself, then sometimes exploding into a 200-foot wall of flame. Enormous convection clouds punched up to 30,000 feet, fed by the fire's own wind. Firestorm winds toppled some trees before the fire consumed them. The Huck Fire began when a tree fell on a power line on the John D. Rockefeller, Jr. Memorial Parkway, closing travel between Grand Teton and Yellowstone national parks. The Flagg Ranch was evacuated, and Grant Village was evacuated for a second time. By the day's end, the fires had grown more than 50 percent and had torched an additional 165,000 acres.

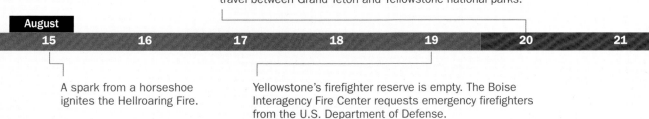

The fires burning in the park double in size on "Black Saturday." Firefighting efforts are halted by extreme winds, exploding fireballs, and massive walls of flame.

A power line struck by a fallen tree ignites the Huck Fire and closes travel between Grand Teton and Yellowstone national parks.

August					
15	16	17	18	19	20	21

A spark from a horseshoe ignites the Hellroaring Fire.

Yellowstone's firefighter reserve is empty. The Boise Interagency Fire Center requests emergency firefighters from the U.S. Department of Defense.

August 22–28, 1988

On August 22, after only two days of training, the first battalion of U.S. Army soldiers arrived in the park to relieve civilian firefighters. They cleared fuels from around developments and mopped up fires. Park concessionaires TW Services and Hamilton Stores evacuated over 800 employees and visitors the next day, closing all facilities except stores at Old Faithful Inn and Mammoth. Visitors continued to traverse the park and were advised to keep all their belongings with them.

On August 24, the Falls Fire joined the Red–Shoshone Fire to form the Snake River Complex. That same day, firefighting personnel directed the evacuation of Canyon Village because of the imminent threat by the North Fork Fire. Firefighters set a large burnout outside of Canyon Village. On August 25, fire managers divided the North Fork Fire into two Incident Commands—the North Fork and Wolf Lake fires—one commander could hand over one-half of the fire, the Wolf Lake Fire made a run toward Canyon Village. The previously set burnout saved the development.

On August 26, amazingly, resources were pulled off fires after a meeting of top brass in West Yellowstone. Incident commanders were directed to cut their losses. No more firelines would be dug except to protect park

Preceding page: A firefighter battles the North Fork Fire on Black Saturday as flames approach the Norris Museum in the Norris Geyser Basin. Photo by Jeff Henry, Roche Jaune Pictures.

Top left: A firefighter watches as fires burn near Grant Village. Photo courtesy of Bob Zellar, The Billings Gazette.

Top right: A firefighter hoists a firehose over her shoulder in the Lower Geyser Basin. Photo by Jeff Henry, Roche Jaune Pictures.

Bottom: Firefighters hose down the roof of the Norris Museum, fearing that fires would shower embers on the shake roof. Photo courtesy of Larry Mayer, The Billings Gazette.

villages, border towns, and structures. People began to realize there was little hope for putting out the fires.

August 29–September 4, 1988

The Mink Fire joined the Snake River Complex on August 29. On August 30, a day that became known as "Grey Tuesday," winds shot balls of fire as far ahead as one and a half miles, driving some fires seven miles downwind and widening other fires by as much as two miles. Fort Lewis Air Force crews began battling the Snake River Complex fires.

By the first day of September, the 109,000-acre North Fork Fire was one and a half miles away from West Yellowstone. The Snake River Complex had grown to more than 156,000 acres. The Clover–Mist Fire was mapped at more than 231,000 acres.

For much of that smoke-filled summer, townspeople in three of the park's border towns—West Yellowstone, Silver Gate, and Cooke City—feared for their livelihoods. From September 2 to September 4, they feared for their property. In West Yellowstone, Montana residents and Idaho neighbors installed an irrigation system around a power substation on September 2. Firefighters also set a backfire to divert the North Fork Fire from West Yellowstone and a nearby development of summer homes.

On September 3, north of the park, thirty-nine firefighters deployed emergency fire shelters during a firestorm on the Storm Creek Fire. They were stationed at the Silver Tip Ranch to protect the historic structures from the Storm Creek Fire, which one firefighter described as "burning like a freight train at 9 P.M." No buildings or lives were lost. On

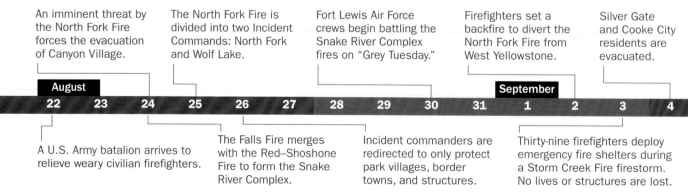

An imminent threat by the North Fork Fire forces the evacuation of Canyon Village.

The North Fork Fire is divided into two Incident Commands: North Fork and Wolf Lake.

Fort Lewis Air Force crews begin battling the Snake River Complex fires on "Grey Tuesday."

Firefighters set a backfire to divert the North Fork Fire from West Yellowstone.

Silver Gate and Cooke City residents are evacuated.

August 22 23 24 25 26 27 28 29 30 31 **September** 1 2 3 4

A U.S. Army batalion arrives to relieve weary civilian firefighters.

The Falls Fire merges with the Red–Shoshone Fire to form the Snake River Complex.

Incident commanders are redirected to only protect park villages, border towns, and structures.

Thirty-nine firefighters deploy emergency fire shelters during a Storm Creek Fire firestorm. No lives or structures are lost.

Above: In the Madison River Valley, cached firefighting tools are laid out in a circle so the firefighters can safely deposit and retrieve them.
Photo by Jeff Henry, Roche Jaune Pictures.

Right: Firefighter Mark Courson gives orders to fire crews working along a bulldozed fire line near Silver Gate.
Photo courtesy of James Woodcock, *The Billings Gazette*.

September 4, residents evacuated Silver Gate and Cooke City, which were also threatened by the Storm Creek Fire, and burnouts were ignited near the park's northeast entrance to protect those towns.

September 5–12, 1988

By September 5, there were 9,600 people fighting fires in Yellowstone and 117 aircraft and eleven of the nation's most experienced fire-management teams supervising thirteen named fires in the Yellowstone area. Governor Ted Schwinden of Montana banned all outdoor recreational activities in the state on September 6.

Early in the morning on September 7, visitors were evacuated from the Old Faithful area, although several were still around when the North Fork Fire burned over the area later in the day. The Old Faithful Inn survived the fire, but nineteen small cabins and buildings were destroyed.

Also on September 7, residents of Silver Gate and Cooke City feared the worst when the backfire that firefighters had purposely set to protect the two towns from the Storm Creek Fire did not work. Sparks from the backfire had ignited a new fire, which was now a more serious threat. At the same time, the Clover–Mist Fire was wreaking havoc near the park's East Entrance. People at Pahaska Tepee and Shoshone Lodge were evacuated. Twenty-five miles north, at Crandall Creek and Squaw Creek in the Clarks Fork

Facing page: Firefighters spray down the Old Faithful Inn as the North Fork Fire blazes though the area on September 7, 1988.
Photo by Jeff Henry, Roche Jaune Pictures.

Valley, the Clover-Mist Fire destroyed or damaged seventeen trailers, five residences, one store, three outbuildings, two vehicles, and two boats.

The next day, President Reagan dispatched cabinet officials—Interior Secretary Donald Hodel, Agriculture Secretary Richard Lyng, and Undersecretary of Defense William Taft—on a fact-finding trip to Yellowstone. They were shocked by what they found. At this point, fires in the Yellowstone ecosystem had burned more than one million acres. They promised more help from the military and a review of Park Service and Forest Service fire management policy. Congress gave Canadian firefighters authorization for quick entry into the United States, for there was no time for border delays.

On September 10, Yellowstone National Park closed to the public for the first time in history. People evacuated ahead of three fire fronts. Residents from Mammoth Hot Springs and Duck Creek (near West Yellowstone) escaped ahead of the North Fork Fire. Jardine, Montana, residents fled from the Hellroaring Fire. Firefighters successfully protected structures at Onemile Creek in the Clarks Fork Valley from the Clover–Mist Fire.

September 11 brought several welcome arrivals. The first snow fell on tired firefighters, slowing the growth of the fires. Local radio stations played "Jingle Bells" and other Christmas carols. And the U.S. Marines arrived to help with the fires. The Huck and Mink fires merged and were managed as the Huck–Mink Fire. On September 12, a pilot died in a light-plane crash near Jackson, Wyoming, after transporting fire personnel.

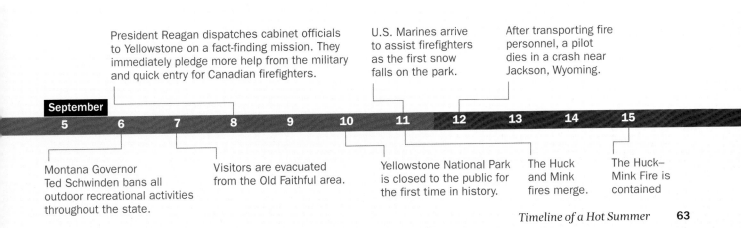

President Reagan dispatches cabinet officials to Yellowstone on a fact-finding mission. They immediately pledge more help from the military and quick entry for Canadian firefighters.

U.S. Marines arrive to assist firefighters as the first snow falls on the park.

After transporting fire personnel, a pilot dies in a crash near Jackson, Wyoming.

September

5   6   7   8   9   10   11   12   13   14   15

Montana Governor Ted Schwinden bans all outdoor recreational activities throughout the state.

Visitors are evacuated from the Old Faithful area.

Yellowstone National Park is closed to the public for the first time in history.

The Huck and Mink fires merge.

The Huck– Mink Fire is contained

September 13–30, 1988

After the break in the weather, firefighters were able to contain the Huck–Mink Fire on September 15. Two days later, the number of military personnel in the park peaked at 4,146. By the end of the month, most of the military personnel had left Yellowstone National Park because cooler weather, higher humidity, and precipitation had virtually stopped fire growth in the ecosystem, and the need for emergency firefighters had waned.

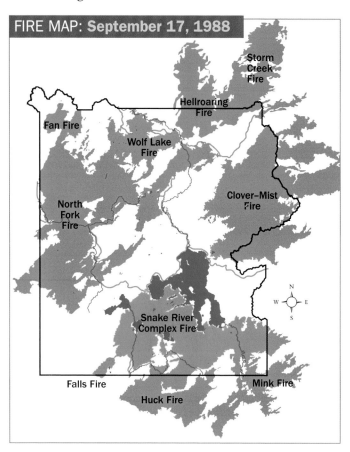

FIRE MAP: September 17, 1988

Storm Creek Fire

Hellroaring Fire

Fan Fire

Wolf Lake Fire

North Fork Fire

Clover–Mist Fire

Snake River Complex Fire

N E S W

Falls Fire

Mink Fire

Huck Fire

October–November 1988

There was very little fire activity in the park after the first snow, though the fire crews were busy mopping up fires until about mid-November. Outside the park, some fires showed increased size, but that may have been because of mapping discrepancies. On October 11, a Bureau of Land Management employee was killed by a falling tree while mopping up on the Clover–Mist Fire in Shoshone National Forest.

There may not have ever been an official declaration of "containment" for most of the fires because, as Vegetation Management Specialist Roy Renkin noted, "Smoke could be observed way out in the black on most fires up until the arrival of over-winter snows and no one wanted to call it over until the fat lady started singing." When all the fires had a few inches of snow on them, most of the fires in the park were declared out on November 13, 1988. On November 18, the Huck–Mink Fire was the last fire to be declared out.

But the story doesn't end there. A column of smoke was reported in June 1989 in Yellowstone's backcountry, near Broad Creek southeast of Tower Falls. Park staff investigated the report and found no evidence of a recent lightning strike, leading them to believe that it was a last remnant of the North Fork Fire, the largest fire that burned in 1988. ◼

Preceding pages: As the North Fork Fire approaches on September 7, 1988, park visitors and employees huddle on a boardwalk near Old Faithful Geyser, which was a relatively safe zone because of the absence of fuel.
Photo by Jeff Henry, Roche Jaune Pictures.

Facing page: National Park Service rangers Les Brunton and Bob Duff cut snags along the West Entrance Road in Madison Valley, in the wake of the North Fork Fire, August 28, 1988.
Photo by Jeff Henry, Roche Jaune Pictures.

Fire crews continue to mop up spot fires. A Bureau of Land Management employee is killed by a falling tree in the Clover–Mist Fire.

The Huck–Mink Fire is the last to be declared extinguished.

October
8 9 10 11 12 13

November
12 13 14 15 16 17 18 19

With winter snow covering most fires, officials declare them extinguished.

Chapter Five

Fire Stories

"Yellowstone is a national treasure. It seemed important at the time that people should be able to see what fire looks like, if they could see it safely. It was smoky and not pretty, but it still was an important part of the learning process."

—DAVE PONCIN, Incident Commander, 1988 Yellowstone fires

An exciting way to tell the story of the 1988 fires in Yellowstone was to ask people who had experienced the fires to share their stories. I discovered that everyone had a story about 1988, be it a vivid memory, a frustration, or a close call.

Stories were relayed to me by fire decision-makers, such as the park superintendent and an incident commander; by "ground-pounders," which is how firefighter Jane Lopez, my first interviewee, described herself; and by those who were in the area, including a neighbor and a park visitor. Looking back, this "fire event" was something that the people I spoke to were glad they had experienced, especially once it was over. They will remember the awesome summer of fire for the rest of their lives.

The Calfee Creek Cabin Incident

Jane Lopez, *Helitack Firefighter*

On July 14, 1988, Jane Lopez was doing backcountry trail work three miles from the Calfee Creek Cabin to prepare for presidential candidate George H. W. Bush's visit. Although she was in Yellowstone that summer to work on a helitack crew, she and crewmates Kristen Cowan and John Dunfee were assigned to work on trails that day, so they took trail packs instead of fire packs, although Dunfee had a fire shelter in his pack. Their supervisor, Dick Bahr, assured them that if they needed anything, he'd fly it in. He didn't anticipate that a nearby fire would intensify and that Bush's trip into Yellowstone from the Shoshone National Forest would be cancelled.

The temperature climbed, and by late afternoon the Clover Fire, a small "natural burn" in the park's northeastern corner, flared up. Lopez heard Chief Ranger Dan Sholly and Bahr talking over the radio she kept on her belt about possibly doing a cabin protection at the Calfee Creek Cabin. Sholly radioed the three-person helitack crew and told them he would pick them up by helicopter to assist in protecting the cabin.

Lopez grew increasingly concerned—she and Cowan didn't have their fire packs and life-saving fire shelters, and although Cowan had basic fire training, she hadn't had any actual fire experience. Lopez called 700 Fox, the radio call number for the Fire Cache, to request their fire packs even before the three firefighters got into the helicopter. After a confusing number of radio conversations, no packs were delivered. Once inside the helicopter, as they were securing their seatbelts and heading toward the cabin, Lopez told Chief Ranger Dan Sholly twice that the two women didn't have their fire shelters.

As pilot Curt Wainwright cautiously circled down toward the cabin, he said he was too heavy due to the high temperatures from the fire and had to let someone out. That someone was Lopez.

"I didn't have a fire shelter, and it was obvious the fire wasn't going to stop!" Lopez said, with passion.

Preceding pages: Exhausted firefighters rest after spending hours building fire lines, the grueling job of clearing trails through the forest in an effort to stop the spread of fire. Photo courtesy of Larry Mayer, *The Billings Gazette*.

Right: Firefighter Larry Williams of San Luis Opispo, California, at Gibbon Meadows August 18, 1988. Photo by Jeff Henry, Roche Jaune Pictures.

After the chopper landed on a ridge, Lopez got out and came around the chopper to get her trail pack from the chopper's other side, but Sholly waved her away. The helicopter rose into the smoky sky, leaving Lopez alone. No water, no pack. Ash was starting to float out from the fire column and fire retardant had been ordered. This is not good, Lopez thought. Fortunately, she had her park radio and she called the pilot, reminding him of her location. She wanted to get out.

Pilot Wainwright dropped Sholly, Cowan, and Dunfee near the Calfee Creek Cabin and waited as long as he could. Wainwright radioed Sholly three times and offered to take them out. Sholly refused; they were there, he said, to protect the cabin. Wainwright then picked up Lopez and landed on another ridge in full view of the fire as it approached the cabin. Lopez said she watched, terrified, as the fire passed over Sholly, Cowan, and Dunfee. Sholly and Cowan huddled under Sholly's fire shelter while Dunfee found refuge in his own.

When it was over, Lopez flew to the Lamar Valley and drove a truck back to the Fire Cache in Mammoth. Dunfee, after his ordeal, was instructed to protect the cabin with Ranger Brian Helms, who was later flown in. The Calfee Creek Cabin survived.

"The Calfee Creek Cabin incident was a hell of a way to start the fire season," said Lopez, who currently works as a fire program manager. "I had to push it to the far reaches of my mind." ☒

Facing page: A smoky crosswalk sign in Yellowstone National Park. Photo by Jeff Henry, Roche Jaune Pictures.

Below: Jane Lopez flying on the helitack crew. Photo by Steve Brigham.

Evacuating Grant Village

Carol Shively, *Park Ranger Naturalist*

Carol Shively, who served as a fire information officer for the Snake River Complex fires, remembered the evacuation of West Thumb on July 31. The Shoshone Fire swept over West Thumb; the visitors Carol Shively evacuated were—as she later described in an article for *Natural History* magazine—"captivated by the mushroom cloud of smoke rising to the north, the helicopters dipping low to fill their buckets in the lake, and the brilliant red retardant drops flashing across the sky [and a] moose and her calf swimming just off the lakeshore." Fire engine crews stood ready to protect West Thumb's buildings, as fire rolled on a wave toward them.

Near West Thumb, Shively said watching the fire crown in the trees around the meadow "safe zone" that surrounded them was extraordinary. She later wrote in the *Natural History* article that she had witnessed "something so extraordinary" that her mind "must stretch to include it."

The east flank of the fire ended its dramatic run at Yellowstone Lake. The historic West Thumb Ranger Station, built in 1925, still stood.

Then, on August 20, Black Saturday, Shively had just finished leading a walk at West Thumb Geyser Basin when she noticed the Snake River Complex fire was making a run at Grant Village. Nearly 100 visitors were stranded at the visitor center. The roads north and south of Grant Village had been closed by the advancing fires. There was no way out.

As the fire approached the Grant Village Campground, the fire camp had to retreat to the relative security of the general store's expansive parking lot. Winds roared, smoke swirled, and several lodgepole pines toppled over, blocking the visitor center parking lot exit, stranding about 100 visitors.

Back at West Thumb, Shively drove through the barricade toward Grant Village. When she arrived at the visitor center, she told the visitors what she'd just witnessed between West Thumb and Grant Village. "We have to get people out of here!" she said.

The window of opportunity presented itself in the next hour. In the chaos, Dan Buss managed an orderly evacuation of the visitor center.

Park ranger naturalists directed the stranded visitors out of the area. Shively; her supervisor Roger Anderson; and another ranger naturalist, Ellen Petrick, ran three abreast to the lodging area of Grant Village. Shively entered the lodge's dining room. "Ladies and gentlemen, put your forks down," she said to the sea of diners seated before her. "You have ten minutes to evacuate. Don't pay your check, get in your cars and go now." She looked at the cashier and the shift manager and said, "You, too, don't worry about the money. Get yourselves to safety."

At the Grant Village Lodge, Anderson spoke with the location manager, and their staff knocked on hotel doors and employee housing units. He commandeered firefighter buses to transport employees without vehicles. The bus drivers initially balked, but the rangers told them the firefighters would stay and the employees would go. They were bused to Fishing Bridge, a safe zone protected by Yellowstone Lake.

For Shively, the 1988 fires were life-changing. "I felt if I could survive the 1988 fires, I could survive about anything," she said. "It was as if we wore an invisible purple heart on our uniforms from that time. The experience deepened my conviction to help teach people about the power and necessity of natural processes in big wild places like Yellowstone." ☒

From Farmer to Volunteer Firefighter

Leon Martindale, *Volunteer Firefighter*

Leon Martindale, a farmer and life-long resident from Ashton, Idaho, was waterskiing on Island Park Reservoir when he saw a column of smoke from the North Fork Fire. It was July 25, and he recalled thinking, boy, they'd better not let that fire go. It was dry. His crops—barley, hay, peas—had already begun to wither. Little did he know, he'd be fighting that fire a month later.

In late August, when the North Fork Fire was threatening West Yellowstone, Mormon church leaders called area wards, and Mormon farmers responded with tractor trailers to haul water pumps and irrigation lines to West Yellowstone. Leon Martindale, Layle Cherry, and

Dale Clark from Ashton, Idaho, kept the diesel-powered water system pumps fueled and operational.

Martindale estimated that nearly 400 people—including townspeople from West Yellowstone, Montana, farmers from Ashton and Rexburg, Idaho, and students from Rexburg College—worked together to save threatened homes and the electrical substation from the advancing fire. To draw water from the Madison River, farmers laid a half mile of main line, which distributed water through eight quarter mile sections of pipe and sprinklers that flanked the southern and western edges of the town.

Townspeople braced themselves as the fire approached. There were skeptics in the beginning. How could these farmers and their irrigation system make a difference? "Farmers can put things together that other people can't," Martindale explained succinctly.

The plan worked. The fire circumnavigated the town, partly because of their efforts.

After the fire passed by the town, Martindale was packing up when the three farmers were called to help protect Montana Power Company's electrical substation and other structures at Old Faithful. They loaded up the pumps and pipe, jumped in their truck, and headed for Old Faithful. This time there would be no beds for the Idahoans; they would doze off and on in their truck.

At Old Faithful, Martindale recalled positioning the "big gun"—a huge sprinkler that shoots water 150 feet—on the substation. Two days later, the firestorm blew over to this area, the park's primary tourist destination. As the firestorm approached, Martindale and his comrades rescued a terrified young employee who was hiding among boxes between buildings behind the Old Faithful Inn. The farmers hauled her and others into the back of their pickup truck and took them to the lodge. The men returned to the area behind the Old Faithful Inn, fire hose in tow, and then crawled up onto building roofs to put out spot fires. The inn itself was protected by the water cascading from its sprinkling system.

During the firestorm's extreme heat, the men stayed in their pickup with the air conditioner on. When the firestorm miraculously passed through the area with minimal damage—and no damage to the beloved inn—Martindale and his friends got out and found that the paint on their truck had changed color.

"Everything all around was burnt as black as could be, except for the half mile of line that had been watered," Martindale said. "There, the grass was green." ☒

From Retiree to Hot Shot Firefighter

Ron Scharfe, Hot Shot Firefighter

Ron Scharfe came to Yellowstone in early August 1988. He had retired from twenty-two years as a Chicago firefighter and was living in Missoula, Montana, when Yellowstone needed firefighters. He was visiting his son, Pat, who was a member of a hot shot crew, when someone phoned Pat to fight the fires. Pat was traveling, so Scharfe volunteered to go in his stead. At the Missoula Smokejumper Base, he passed the firefighters' arduous step test to earn his "red card," which certified he was physically able to fight fires on federal lands. When the tester asked his age, Scharfe replied, "Would you believe fifty-seven?" Even though he had to show his driver's license, which revealed his actual age of sixty-one, Scharfe still walked away with his red card.

He remembered the day Canyon Village was nearly engulfed by flame. "The smoke was terrible; you could see maybe fifteen feet in front of you," Sharfe said. He said that soldiers from Fort Lewis were digging fire lines, the buildings were all foamed down, and the North Fork Fire was fast approaching the developed area.

"This place is gonna be history in thirty minutes," Scharfe recalls telling the fire crew boss in charge of protecting structures. Scharfe quickly related how he thought the structure at Canyon Village could be saved. Scharfe was handed the bull horn to set up the defense strategy.

Scharfe positioned approximately thirty firefight-

Facing page: Near Mammoth Hot Springs, flames from the North Fork Fire torch a Douglas fir tree and light up the night sky.

Photo by Jeff Henry, Roche Jaune Pictures.

ers, each equipped with a water hose, in front of the Canyon Visitor Center. He split them into two groups: one stood on the left side of the building, the other on the right. He directed them to let loose cooperative, interlocking streams of water to form a water curtain in front of the building. The curtain became an ephemeral screen of fog, cooling the air, and the fire blew around the building.

After the close call was turned back using Scharfe's strategy, the fire crew boss asked what he could do for Scharfe. Scharfe replied, "Put me on the hottest hot shot crew in the park." For the rest of the summer, he served on the Wyoming Hot Shot crew in Yellowstone's backcountry. He was interviewed on the television program 20/20, because the reporters were so impressed with his story.

Scharfe believed that visitors should not have been allowed into the park during the firefighting in 1988. He also thought that the National Park Service and the Forest Service needed to "get together" and cooperate more fully. He used the example of the Park Service threatening to fine the Forest Service $500 because one of their buses was parked on the grass to pick up firefighters.

One of the best parts of that summer, Scharfe said, was that he got to fly into places where other visitors don't normally go. At one point during the summer, he and two compatriots were evacuated by helicopter during a big burnover. They saw a grizzly bear burying a buffalo. As they flew about twenty-five feet above the scene of the bear and its treasure, the bruin stood up to look at them. That once-in-a-lifetime image was etched into Scharfe's brain. ⊠

Frustrations on the Fireline

Dave Poncin, *Incident Commander*

Incident Commander Dave Poncin's toughest moment was on August 15, 1988, when the North Fork Fire charged down off the Madison Plateau, crossed the Firehole River, and leapt across the road at Madison Junction. Poncin and his crew had hoped the road between Old Faithful and Madison would serve as a

fire line. It didn't. Flames jumped the road, sending out spot fires half to three-quarters of a mile ahead so quickly that firefighters couldn't put them out. The fire climbed into canyon country, a broken country with different wind patterns that spelled trouble. The North Fork Fire seemed to be resisting control at every turn. Ten days later, the fire had spread so fast that it became too large for one Incident Command and was split into two fires: North Fork and Wolf Lake.

Nearly a month earlier, Poncin had been asked by the Park Service to manage the Snake River Complex fires. As one of two Type I Incident Commanders in the Rocky Mountains and a former Gallatin National Forest district ranger, Poncin knew the country, the wildlife, and how to work with the Park Service.

When he arrived on July 24, 1988, Poncin successfully directed his crews to execute a "burn out" at Grant Village using a helicopter and ground crews. To protect the development's power station, he bulldozed the first fire line in the park that summer. This victory was a much-needed morale booster.

Poncin recalled that after about mid-August, freakish fire behavior was stirred up by the dry cold fronts that moved through the park, whipping up multidirectional winds that made firefighting extremely frustrating and difficult. Poncin's team had never before had the responsibility to protect historic structures and wooden bridges, as they did during 1988.

Poncin feels that Park Superintendent Bob Barbee made the right decision to keep the park open to visitors—Poncin and his team complied by trying to keep roads open, although they were closed intermittently that summer.

"Yellowstone is a national treasure," Poncin said. "It seemed important at the time that people should be able to see what fire looks like, if they could see it safely. It was smoky and not pretty, but it still was an important part of the learning process."

Poncin also saw the media coverage as beneficial, although he had to attend numerous public meetings in West Yellowstone and deal with the persistent

Right: Incident Commander Dave Poncin at the Madison fire camp.
Photo by Jeff Henry, Roche Jaune Pictures.

presence of forty to fifty media people in the Madison fire camp. "The public had a greater understanding of the park than at any other time," he said. "The 1988 fires were experienced by more people than any other event in Yellowstone's history."

Poncin said he is grateful to have witnessed the 1988 Yellowstone fires, pointing out that many good changes came about as result of the fires, including renewed forests, increased public fire education, and better fire management.

"Put the 1988 fires up against natural calamities like volcanic activity and erosion of the Grand Canyon, and fires really don't seem like much," he said. "The things that we consider so beautiful today had to have been a disaster when they happened."

Below: Firefighter Rick Hutchinson and Bonnie Gafney place foil fire shelters on the roof of the Nez Perce Patrol Cabin.
Photo by Jeff Henry, Roche Jaune Pictures.

Facing page: Tourists race to beat the fire that was threatening to jump the road south of Norris. Roads closed by fire often disrupted tourist traffic.
Photo courtesy of Larry Mayer, *The Billings Gazette*.

Saving Backcountry Cabins

Phil Perkins, *Fire Management Officer*

When the North Fork Fire made a run for Mammoth in early September 1988, Phil Perkins's wife Debra and their three children and other evacuees found refuge at a Holiday Inn in Bozeman. When they walked in, the receptionist sniffed the air. "You've just evacuated from Mammoth," the receptionist said to Debra. The smell of smoke permeated all of their clothes.

Dad was at the other end of the park, saving a backcountry cabin.

After Perkins and ranger Nick Herring had successfully protected the Buffalo Plateau Cabin on August 31, eliminating fuels that would bring fire lapping to the cabin's edge, they were sent to the Harebell Cabin along the park's southern border to do the same. They kept the area around the cabin watered with two water pumps, overlapping water lines, and sprinklers. Over the next three to four mornings, the

men awoke at approximately 2:00 A.M. because the Huck Fire made middle-of-the-night runs on the cabin from different directions. During one long hour, the men hunkered down on the leeward side of the cabin because of unbearable heat on the other side.

District Ranger Gerry Mernin wanted them to save the grass for his horses, but little remained. But Perkins and Herring had saved the cabin. Herring didn't have much experience fighting fire, and Perkins didn't let on until afterward that "it was kinda tight."

Perkins, who now serves on a Type I overhead fire management team for the Northern Rockies—an esteemed position in the world of firefighting—began his career in Yellowstone National Park in 1985 as an assistant fire management officer. In 1988, he managed fire monitoring, dispatch communication, lookouts, helicopters, delivery of food and water to firefighters, and transitions between firefighting commands. During the times when the Park Service

resumed "control" of the fires from Incident Command teams, "everybody did everything."

August 20 was a day that Perkins deviated from his usual routine. He had been given a rare day off—ironically, that day made history as "Black Saturday." Instead of taking free time, Perkins hiked the Thunderer, a 10,000-foot ridgeline, with forty firefighters, two crew bosses, and Chief Ranger Dan Sholly in a last-ditch attempt to keep the Clover–Mist Fire from whipping down into the Soda Butte drainage and threatening Silver Gate and Cooke City.

They started hiking about dark on Saturday and arrived at the ridgetop before midnight. They worked through the night. Early the next morning, Perkins remembered three individual convection clouds "punching up" from the fires—he had never seen fire activity like that so early. Winds whipped to 40 mph by noon in the valley and up to 80 mph on the ridges. They were evacuated by helicopter at midday. His so-called "day off" was etched forever into his memory.

Perkins said there were communication problems among the agencies during the fire season of 1988—absent was a guiding document that spelled out the forests' and parks' fire policies. Sensational media pieces didn't help. Communication greatly improved in 1989 after one interagency plan was written, and new fire management positions were created. ☒

Postcards from Yellowstone: August 1988

Frank Markley, *Park Visitor*

Frank Markley had reservations to stay at Old Faithful Inn for a week in late August to early September 1988. As his vacation date drew near, the television was broadcasting gloomy stories of fire-threatened Yellowstone, and Old Faithful in particular. Nevertheless, Markley left Ohio, headed for Sturgis, South Dakota, then he continued west toward Yellowstone. He thought this would undoubtedly be an experience of a lifetime, no matter how it turned out.

As he drove west, Markley listened to hourly radio updates. When he got to the East Entrance on August 30, the ranger on duty didn't charge him admission "because of the fires." She handed him a printout with visitor precautions on one side and a letter signed by Park Superintendent Bob Barbee on the other. He tuned in to a radio station for fire updates.

Travel around the park was a challenge. Visitor centers and park rangers kept visitors informed as best they could, but conditions changed quickly, closing roads. Markley was told Canyon was closed, so he decided to visit Norris. By the time he arrived, Norris, too, was closed. The next day, he found the road to Artist Point closed again, and instead visited Mud Volcano, where three busloads of firefighters were laughing and joking as they walked around. Markley would finally glimpse the Grand Canyon of the Yellowstone three days later.

The following day, Markley encountered the thickest smoke he'd seen so far, at the South Entrance. On his way to Grand Teton National Park, he had to keep his headlights on to penetrate the smoke. While at the Moose Visitor Center, he heard the South Entrance

had opened again—though he didn't know it had closed—so he promptly headed back to Old Faithful.

Markley's final two days were spent at Old Faithful—a welcome change from trying to negotiate park roads and their closures. On September 6, a fire alarm went off inside the Old Faithful Inn and staff rushed to close the heavy metal double doors at the ends of the halls, something he'd not noticed on previous visits. The power also blinked out at Hamilton's Upper Store, changing his lunch plans. At the other store in the area, Markley overheard one waitress say to another, "I'm taking bets on the day and time," referring to their inevitable evacuation. Markley strode around the smoke-filled geyser basin, the familiar far side view of the inn obscured.

Frank Markley got up early the morning of September 7. His room in the "Old House" of the inn was very hot, and it didn't help to open windows. When he checked out before 6:00 A.M., he was surprised to see several young employees hanging around the front desk. A half hour later, as he was driving north, the radio announced that Old Faithful was being evacuated. As he drove through the Roosevelt Arch near Gardiner, Montana, it was the first time he felt relief at leaving the park. Though smoke and the smell of Yellowstone burning followed him through three states as he motored toward his Ohio home, Markley was grateful to have experienced the fires of 1988.

Markley's affair with Yellowstone has continued, with regular visits. He had wondered if the park would recover in his lifetime, but after his 2007 visit, he said it was "uplifting to see the old snags falling over and the bright green hillsides . . . Frankly, I think it's happening faster than I thought it would." ☒

Saving Old Faithful

Jeff Henry, *National Park Service Photographer*

Excerpted and condensed from *Old Faithful Inn: Crown Jewel of National Park Lodges* by Karen Wildung Reinhart and Jeff Henry

The Old Faithful area was initially threatened by the North Fork Fire shortly after a woodcutter in Idaho's Targhee National Forest started the blaze with a

carelessly discarded cigarette on July 22. The second threat to Old Faithful started the afternoon of September 5, when the wind from the east and northeast pushed the North Fork Fire past the fire lines. A run was underway and I was there to photograph it—when it became apparent in late July that Yellowstone was on the threshold of an exceptional fire season, the National Park Service assigned me to photograph the fires for the park archives.

I found two firefighters on the roof of Old Faithful on September 7, who used infrared scopes to identify hot spots through the smoke. They were scanning the timbered ridges west of Old Faithful. Slurry bombers were dropping their loads of red fire retardant. Several helicopters were flying about as well, carrying observers or carrying buckets to dump water on the fire. Soon after I arrived on the Inn's roof, the winds picked up to gale force. I've heard variously that the winds were 50 to 80 miles per hour that afternoon at Old Faithful.

By September 7, the North Fork Fire had been burning for several weeks and the winds had made it explosive. Behind the smoke was a blitzkrieg of fire and wind that was bearing down on the Old Faithful area. Although the flames were shrouded by smoke, once in a while, it would clear and I could see flames three to five times the height of the 100-foot lodgepole pines. Usually the flame presented itself as a wall, a long curtain of fire that advanced like a line of soldiers in formation. But sometimes huge balls of flame flew from the wall of fire, as the wind caught a pocket of gasified fuel and flung it forward. Some fireballs were several hundred yards in diameter and were hurled several hundred yards ahead, where they ignited new pockets of fire.

The fire grew stronger as it advanced, a gigantic rolling wave of flame, sucking in air along the ground, while the wind blew a crest of flame forward. Like an ocean wave breaking, the flame rolled forward and curled under itself at its base. I saw spirals of fire like flaming tornadoes. The one constant was the fire's freight-train-like roar.

By now, we didn't need any infrared scopes to see

what was coming. The people on the roof of the Inn beat a hasty retreat. Just the two firefighters and I remained on the widow's walk. The two of them donned fire protective masks and gloves, agreeing that nothing could be done about what they thought was an inevitable catastrophe. I saw no reason to disagree. It looked as though the area buildings would be torched in a firestorm and a lot of people might die.

In the parking lot, I was struck by a blizzard of sparks and embers as a billow of smoke surged just overhead and light nearly vanished. There were still many visitors in the area, and I imagined taking one or two families by the hand and leading them to the geyser plains in front of the Inn, where I had further visions of squatting in the water of the Firehole River while we watched the Old Faithful Inn and other buildings burn. That seemed the best I could do.

Fire had already entered the Old Faithful complex—I could hear cabins behind the Snow Lodge exploding, like the artillery bombardment in a war zone. Indeed, the very air seemed to be on fire.

Back at the Inn, a large number of firefighters gathered, holding hoses, and firefighters with water-pumping trucks were wetting down the sides of the building as high as their hoses would spray. The Inn's sprinkler system was on and water deluged the eaves. Those firefighters, some of whom had even stationed themselves perilously on the flat roofs of the East and West wings of the Inn, saved the historic structure. Without their brave efforts to keep the walls and roofs of the Inn wet, the building certainly would have ignited that afternoon.

Several times that day, I saw upturned stumps in the geyser basins or fallen logs on islands in the parking lot burst into flames. A ground blizzard of sparks and embers swirled around my ankles and calves. Larger embers showered down upon the Inn like fiery ejecta from an erupting volcano. One of the Inn's outbuildings did catch fire that day, but it was quickly put out by some alert concessions employees who happened to be nearby.

Then someone shouted that there was fire on Observation Point, above Geyser Hill on the other side of the Firehole River from Old Faithful Inn. It

was true—windborne fire had leapt completely across the Old Faithful area, including all the parking lots and bare geyser plains, and ignited new fire on the basin's other side. The Old Faithful Inn was literally surrounded by fire.

Firefighters continued hosing the Inn, spot fires burned on the southern flank of the Old Faithful development, and spot fires appeared around the Inn and Old Faithful. Gradually, however, as the afternoon wound down and fire moved east, it became safe to believe that the Old Faithful Inn and the other major buildings of the area were going to survive. No one had been killed or seriously injured in the afternoon's conflagration.

I circulated around the Old Faithful area that evening, looking around to see what had been burned, feeling thankful for what hadn't burned. Some vigilant firefighters were still on duty, standing guard through the smoky night while spot fires snapped and crackled in the distance. When I left, late that night, I sat in the cab of my big Park Service truck with its load of firewood and looked at the Inn for one last time. The Old House built in 1903 loomed large in the smoky darkness, against a backdrop of glowing spots of fire over on Observation Point, and reminded me of Sir Christopher Wren's famous photograph of St. Paul's Cathedral standing defiant amidst the smoke and dust of the London Blitz of 1940. I'm sure Londoners couldn't have been more thankful that their cathedral had survived the Nazi bombing, than I was that the Old Faithful Inn had survived the North Fork Fire. ☒

Fire Lessons in 1988

Don Despain, *Research Ecologist*

Research ecologist Don Despain had a front row seat to the natural show that played out on the stage of the Greater Yellowstone Ecosystem in 1988. The fires were the onstage players, fueled by wind, dry conditions,

and high temperatures. Seated next to him were other decision-makers, such as Park Superintendent Bob Barbee, as well as a host of firefighters, visitors, area residents, and the media that looked to Despain and Barbee for guidance, hope, and answers. For Despain, who has since retired but continues to live in the Greater Yellowstone Ecosystem, the 1988 fires were a fascinating natural phenomenon, and he had a ringside seat. He felt fortunate to have that seat. A fire of that magnitude, he said, wouldn't happen again for three or four or five generations.

Despain wasn't worried about the 1988 fires because he had witnessed the earlier Yellowstone fires of 1976 and 1981, and he knew the importance of fire to the ecology of the Greater Yellowstone Ecosystem. Yellowstone wasn't being destroyed and the plants would grow again. And Despain, of course, was right. The forests are growing back. When Despain tried to convey this knowledge to the American public, he was accused of "trying to put a nice face on a terrible thing." Few listened to his ecological viewpoint.

Then came the August 28, 1988, headline in the *Denver Post*: "A good year...'burn, burn,' says fire-loving Yellowstone biologist" above an article written by Jim Carrier. The reporter and Despain had hiked out to a research plot that Despain had staked out ahead of the fire to study postfire regrowth. When fire engulfed his plot—the first plot in his career to burn—Despain was excited. Giddy about his fire study in microcosm, he cried, "burn, baby, burn," not realizing this might sound like a cavalier attitude about the whole park. Even though Carrier put the quote in context inside the article, newspaper readers reacted negatively to the headline. Despain regretted saying those words, which caused others so much trouble.

Looking back through the smoky curtain of controversy, Despain still doesn't think about being a target of public anger. He wasn't perturbed about what people said then, confident in his belief that natural fire had a place in the ecosystem. He said, "People don't understand fire enough to make statements about controlling fire."

Despain evacuated his family from Mammoth by September 10, as did other park employees. He wasn't worried about the fire burning his residence—the fuels that could carry the fire weren't close to Mammoth. He was more worried about his children getting run over by one of the many fire trucks in the neighborhood as the North Fork Fire made its run toward park headquarters. His wife and children fled to Mol Heron Creek, north of Gardiner. A day or so later, they were evacuated again. This time they left the ecosystem and went to Lovell, Wyoming.

Like the rest of America, his children watched newscasts about the 1988 fires, and when their dad started to "fuss and fume" at the television that those Yellowstone acres weren't destroyed and that people couldn't put the fires out anyway, they'd moan, "Here goes Dad again." ☒

In the Hot Seat

Bob Barbee, *Park Superintendent*

When Bob Barbee arrived in Yellowstone National Park in 1983, he said the grizzly bear population was "swirling down a black hole," the hotels were falling down, the bison brucellosis issue was surfacing, and the Park Service decision to remove the Fishing Bridge campground was a hot topic. But in his twelve years as Yellowstone National Park's superintendent, Barbee said the biggest controversy he faced was the fires of 1988. They were, he said, a "character-building experience."

During the stormy summer of 1988, Barbee was the lightning rod for media and others who criticized park policy. "Flawed policy had nothing to do with the fires of 1988—the fires either started by lightning or by humans outside the park," he said. Barbee withstood the criticisms, and fortunately, before arriving in Yellowstone, had learned not to take such things personally. After writing his master's thesis on the 1963 Leopold Report, which condoned natural fire management, Barbee led prescribed burning in Yosemite National Park.

Facing page: Yellowstone Superintendent Bob Barbee at the Old Faithful area with U.S. Secretary of the Interior Don Hodel, during a review of firefighting equipment, July 27, 1988.
Photo by Jeff Henry, Roche Jaune Pictures.

He had been in tough situations before. Many people wanted someone to blame for an event they didn't understand. Someone spelled out their feelings on a West Yellowstone motel sign, "Welcome to the Barbee-que."

The media went crazy, staging high drama in the park, as there was nothing else going on in the country. Presidential candidate Michael Dukakis came to Yellowstone in mid-September, apparently to capitalize on the media presence in the park. When Barbee asked him why he had come, Dukakis replied that it was the only game in town.

Were there things Barbee would have done differently? Sure. Had he known the outcome, he would have taken earlier action on the Fan, Clover, Mist, and Red fires. But average summer rainfall was predicted, and Barbee and his team decided to initially let some fires burn. When it became apparent that the summer weather was unusual, he and others decided to suppress all fires. People forgot that all fires were fought as of mid-July.

After the fires had gone out in late fall and the firefighters had gone home, Barbee traveled to Europe and met with travel media to convince them that Yellowstone wasn't completely burned up, but was still a viable tourist destination. He also had to educate the public about why it was unnecessary to reseed the park; the lodgepole "seed rain," the natural release of seeds from resin-covered cones under fire's heat, would more than adequately reseed the forests of Yellowstone National Park.

Barbee stressed, "This is what is really important: Yellowstone fires were a catalyst to turn the corner on professional firefighting. Communication between the National Park Service and the Forest Service has greatly improved. Trigger points for fire suppression have been identified and technology has evolved."

Bob Barbee predicts that fires in Yellowstone will happen again, the only questions being when and at what magnitude. Just as they did in 1988, fires will occur when all the variables come together—like a symphony composed of just the right players. ▣

Chapter Six

Myth and Science: Toward a New Yellowstone

SIX

"Fire is plural, not singular. One can accurately speak about fire only in conjunction with something else—fire and flora, fire and fauna, fire and earth, fire and water."

—STEPHEN J. PYNE, *Fire in America—A Cultural History of Wildland and Rural Fire*, 1997

I n the aftermath of the 1988 Yellowstone fires, scientists flocked to the park. This was a rare opportunity—never before had there been a living laboratory where they could study the effects of fire on such a scale. Scientists conducted more than 250 studies to investigate questions about fire. The National Park Service spent six million dollars to support thirty-two of these research projects. Plant ecologist Don Despain said the amount and breadth of this research was the most important outcome of the 1988 Yellowstone fires because it gave ecologists a unique opportunity to examine the natural processes of fire.

Preceding pages: Lodgepole pine saplings thrive, sixteen years after the North Fork Fire swept through the Firehole River Canyon.
Photo by Jeff Henry, Roche Jaune Pictures.

Facing page: Sun shines through the frosted limbs of a tree near the Grand Canyon of the Yellowstone River on a very cold winter morning.
Photo by Jeff Henry, Roche Jaune Pictures.

Below: This satellite image shows massive smoke plumes drifting across northern Wyoming and southern Montana. Dated September 7, 1988, this image was taken on a day when high winds fanned the park fires.
Image courtesy of the U.S. Geological Survey, EROS Center.

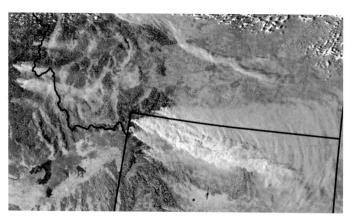

The world's first national park belongs to everyone. People had strong opinions about its future, from scientists to gateway community members, from park employees to the American public at large. Predictions about the effects of the fire on the park ran the gamut from "any land this black will never be green again" to "wildlife will die because there isn't any food." In the following section, the myths or predictions of 1988 are presented and the scientific findings follow.

Myths and Science

Myth: The magnitude of the 1988 fires was unnatural and hadn't happened before.

Science: The history of fire in Yellowstone country started long ago. To understand the severity of the 1988 fires, scientists analyzed fire history, trying to find out whether that epic summer blazed beyond the boundaries of normality. The old volumes of written history only went back to the early 1800s and didn't provide answers.

But Mother Nature had scribed her own stories in wood, earth, and stone.

By coring live trees, examining fire-blackened tree scars on the map of the tree's life, and painstakingly counting annual growth rings, scientists determined that major fires—quite likely as large as the fires of 1988—occurred in the early to mid-1700s. Further evidence was etched in stone; petrified trees exhibited fossilized burn scars.

Scientists then looked at mud flows at the mouths of streams. Over the last 10,000 years, fire debris traveled down tributary streams and deposited clues in

alluvial fans. Fires were more frequent during warm, droughty periods from A.D. 900 to A.D. 1300. Fire was less prevalent during cool periods, such as the Little Ice Age that lasted from A.D. 1500 to A.D. 1850.

Earth science researchers Cathy Whitlock and Sarah Millspaugh found that fire had played a part in Yellowstone ecology for 17,000 years. They gathered sediment core samples from two lakes and read the samples like pages in a history book; pollen from reproducing plants and charcoal from burned trees had laid down annual records of growth and fire. The two researchers discovered that fires the size of those that occurred in 1988 were not unusual. Major fires occurred about every 100 to 300 years and played a vital role in the natural ecosystem.

Myth: Yellowstone's forests will have to be reseeded.

Science: There were no trees planted in Yellowstone after the 1988 fires. Although lodgepole reseeding was discussed that fall, and there was public pressure to do so, ecologists knew there was no need. Ecologists and naturalists had to educate visitors about the role of fire in lodgepole forests.

The forests of Yellowstone's high-elevation central plateau are almost uniformly lodgepole pine, a species that is uniquely adapted to fire. Many lodgepole pines have serotinous cones, which means that they are sealed by resin until a hot fire melts their

Facing page: A rejuvenated forest along the Madison River in 2003.
Photo by Jeff Henry, Roche Jaune Pictures.

Below: The serotinous cones of a lodgepole pine.
Photo by Jeff Henry, Roche Jaune Pictures.

protective coating, sending an explosion of seeds to the forest floor. The spring following a fire, tiny seedlings begin growing, sheltered by the dead trees. Fire is necessary for the rebirth of lodgepole pine.

A visitor to the park after the fires can see that the density and height of the new forests vary. In some places, the trees are so thick and the crosshatch of fallen logs so intricate, that hiking through them is a chore at best. Nearly every year since the fires, researcher John Burger measures the height of lodgepole saplings. He has found that the tallest lodgepoles grow as much as a foot or more per year in open

locations. Twenty years after the fire, some of the park's trees are well over twenty feet tall.

Yellowstone's forests continue to evolve. Eight years after the fires, some new lodgepoles began producing nonserotinous cones, which don't require fire to open the cones. Eventually, the oldest lodgepole pines will tower above the forest floor and the forest canopy will begin to close. Whitebark pine, Engelmann spruce, and subalpine fir, which rely on animals and wind to disperse their seeds, will begin growing beneath the lofty lodgepole pines. Douglas fir trees will grow in low–elevation forests.

Myth: Yellowstone will never be the same. Any land this black will never be green again.

Science: After twenty years of watching green grass, green plants, and green seedlings grow in the park, these fears seemed groundless. Grass, after all, sprouted in burned areas only weeks after the fire. The following year, green growth was abundant, and the carpet of wildflowers was like a rainbow after the storm. By 2006, the forests noticeably tipped the visual scale from black to green. In many areas, most of the charred snags that had towered over the young saplings had fallen, giving way to the young, tall, verdant lodgepole pines. A young visitor might scarcely see signs of the 1988 fires, even though some burned trees will remain standing for fifty years or more.

"Yellowstone will never be the same" was a quote often heard after the 1988 fires. This is partly true, at least during our lifetime. But the scenic landscapes of Yellowstone have burned many times since the glaciers retreated, and these forests many times looked very similar to the recently burned forests of today. The Greater Yellowstone Ecosystem will continue to evolve with fire, as it has for thousands of years. Species that lived in Yellowstone National Park before the 1988 fires have already adapted to fire—their continued presence is proof.

Myth: After the fires, increased erosion will fill rivers with silt and excessive nitrogen, killing the fish. Conversely, small-stream fish populations will increase because of the excessive nitrogen.

Science: There were a few mudslides in the Grand Canyon of the Yellowstone, as well as Lava Creek and Gibbon canyons—steep areas at the caldera's edge—and also in Slough Creek Canyon. In places where the fire burned hot, there was little to stop the soil from slipping downslope and engulfing a stream. Fortunately, the snow melted at a slow and steady rate in 1989, so the increase of silt in the streams wasn't extreme.

Fishing was a challenge for a short time during the summer following the fires, particularly after rainstorms. Nevertheless, the park's best-known trout, the native Yellowstone cutthroat, continued to spawn in their old spawning streams. And, as burned trees continued to topple into streams, fisheries habitat improved.

Most of the seed banks and plant roots in Yellowstone survived the fires—only one-tenth of 1 percent was destroyed. Surviving seeds and roots soaked up nitrogen, Mother Nature's free fertilizer after a fire, and for a few years, the surviving and newly sprouted plants were especially exuberant in their foliage and flower display. Wildflowers sprouted that hadn't been seen in Yellowstone for years. Thick carpets of moss grew in some burned areas and helped

check erosion. The loss of nutrients from land to water was minimal. Fish populations remained steady.

Myth: All ungulates (hoofed animals) will become more numerous.

Science: How did the fires affect other animals, such as bison, mule deer, pronghorn antelope, bighorn sheep, and mountain goats? Researchers found the fires had very little effect on ungulates, except perhaps moose. (See following section.) Scientists found that the numbers of bison, mule deer, and pronghorn were more closely tied to climate and harsh winters. Even though their winter ranges didn't burn, mule deer populations decreased nearly 20 percent after the fires, and pronghorns decreased nearly 30 percent. Very little is known about the bighorn sheep and mountain goats that summer in high alpine areas.

because it gave seedlings and saplings time to grow and increase fire risk. After a few more years, the understory vegetation would grow tall, providing a ladder for fires to reach the crowns of the trees.

So, if fires don't cause lodgepole pine trees to become infested with mountain pine beetles, and beetles don't trigger fires, what is the cause of these infestations? The answer is drought. Drought-stressed trees cannot defend themselves against insects by pitching them out, even if the insects are native. Older trees are more susceptible to drought.

Myth: Rivers that begin in Yellowstone country will flood downstream beyond the park.

Science: The Yellowstone River drains Yellowstone Lake, which is fed by 124 tributary streams. The river, which runs 671 miles north and east from Fishing Bridge to its confluence with the Missouri River, flooded in 1996 and again in 1997, the largest flood years in recent times.

Was this a delayed response to the 1988 fires? Hydrologist Phil Farnes tried to answer this question by comparing the river's peak flow data that had been collected since 1911 with data collected during 1996 and 1997. Twenty-five percent of the drainages above his test site near Corwin Springs, Montana (seven miles downstream from the park), had burned in the 1988 fires. Farnes surmised that the 1988 fires added only about an inch to the peak flow in 1997, and only for one day—an insignificant amount compared to the total runoff volume. If the floods weren't caused by increased erosion due to the burned forests, what was the source of the flooding those two years? Farnes attributed the floods to high winter snowpack.

Myth: Non-native plants will invade areas that burned or were scraped clean during firebreak construction.

Science: Each summer, resource management staff eradicates non-native plants, such as spotted knapweed, yellow hawkweed, and woolly mullein, in developed areas and along roads. They also survey backcountry areas for weeds and have noted an in-creased backcountry presence of only one species: the non-native Canada thistle, an "exotic" that they feel is impossible to eliminate.

Myth: Yellowstone burned because of past fire suppression and the build-up of understory fuels.

Science: Researchers have found that putting out fires in the past may have caused an unnatural build-up of forest fuels in lower-elevation forests that burn frequently, but this is not the case in Yellowstone's high-elevation forests. Because of the cold climate and short growing season, it takes about 200 years for Yellowstone's lodgepole pine forests to accumulate the amount and types of fuels to again support an active forest fire. Prior to the 1988 fires, people had been fighting fire effectively for only about fifty years when helicopters and smokejumpers began to be used for initial attack—a much shorter period of time than the natural rhythm of fire occurrence in Yellowstone. Therefore, the build-up of fuels that did occur because of firefighting efforts simply wasn't enough to cause the fires of 1988.

Myth: Dead trees—both standing and prone on the forest floor—will provide fuel for another huge fire in the near future.

Science: After the fires, in some areas, dead trees stretched as far as one could see. Of course, there were pockets of green trees within burned areas, but the general impression was one of fire-carved, black-ened skeleton-trees everywhere. Understandably, many people feared that the vertical and horizontal maze of partially spent wood would be fuel for an even larger fire—the next time a dry lightning storm flashed over the park.

However, dead trees in a recently burned forest don't carry fire in the same way as living plants and other small-diameter fuels in an unburned forest. Recently burned areas rarely burn again; when they do, the fire burns hot, but moves slowly. A hot, stand-replacing or "crown" fire needs understory "ladder fuels" that move the fire up from the ground into the trees.

Slow-growing spruce and fir retain their tangled lower branches, providing a way for fire to climb to the taller, mature lodgepole pines. Once these understory

Right: Fireweed blooms abundantly among the charred snags.
Photo by Jeff Henry, Roche Jaune Pictures.

fuels burn, subsequent fires usually halt or slow at the old burns. The fire season of 1988 was an exception. In the last month of the fire season, as Renkin said, "It didn't matter what you were, you burned."

Myth: The National Park Service will use prescribed burns so large fires won't occur again.

Science: Prescribed burns, or fires deliberately set by managers under predetermined conditions, have not been widely used in Yellowstone National Park since the 1988 fires. In fact, many other national parks that used prescribed burns as a tool stopped using them after the 1988 Yellowstone conflagration.

Scientists estimated that 50,000 acres would have to be burned each summer to make a difference in the park, and this wouldn't be popular or practical in August, which is the only feasible time to burn but is at the height of visitation in Yellowstone.

Though some forest managers in Yellowstone have considered using fire to eliminate hazardous fuels and create fire breaks, Renkin is cautious.

"We're walking a fine line with the use of prescribed fire," said Renkin. "On one hand, we need very dry conditions to use prescribed fire as a tool to reduce forest fuels, and such fires may be difficult to control. On the other hand, igniting a fire under wet, more controllable conditions won't achieve the goal of reducing fuels."

Myth: The grizzly bear population will decline because many whitebark pine trees burned in 1988. Also, fire plays a role in eradicating whitebark pine from the landscape.

Science: About 24 percent of Yellowstone's whitebark pine trees were consumed by flames; it will take new

trees 60 to 100 years to produce cones. The nutritious seeds in whitebark pine cones are a critical fall food for grizzlies. Though bears were less apt to wander into burned areas looking for whitebark pine cones in red squirrel middens or "cone cellars" during the first five years after the fires, the grizzly population continued to rise. In 2007, they were delisted from the Endangered Species List. Reinhart credits public education as one of the reasons that the grizzly bear population has increased. People, she says, have learned how to live with bears, using bear spray instead of guns and keeping garbage and human foods inaccessible.

Researchers have found that fire does play a role in reducing whitebark pine from the park's high country, but fire is not the only cause of their possible decline. Although new trees sprout from seeds that are cached by birds and animals in both unburned and burned areas, mature whitebark pine stands that have been injured by fire are more susceptible to mountain pine beetle infestation, which can kill the trees. Other threats include white pine blister rust, and some scientists contend that climate change may encourage lower-elevation trees like lodgepole pine to encroach into the whitebark pine zone. Reinhart says, "The bears aren't threatened, their foods are."

Myth: The fires will adversely affect park visitation, even in the long term.

Science: In truth, visitors flocked to the park in the fall of 1988, while the fires were still smoldering under fresh snow. They wanted to see firsthand what was left of their park. Visitation was up nearly 40 percent in October 1988. In 1989, a record 2.7 million people visited Yellowstone National Park, and visitation continued climbing in the years after that.

Some people were surprised at how the fire left varied evidence of its intensity—mosaic patterns of green copses of trees could be glimpsed next to and within burned forests. In some areas, green grass had awakened just a few weeks after fire had burned everything in its wake. In the spring of 1989, bright green lodgepole pine seedlings began to emerge through the baked soil, and acres of pink fireweed and blue lupine swayed in the Yellowstone breeze, anchoring the nitrogen-rich soil and providing stunning opportunities for camera-happy park visitors and professional photographers, offering park visitors and employees insight into nature's amazing forces of renewal.

Surprises after the Fires

Aspen Everywhere!

Research biologist Don Despain remembered that in the spring of 1989 the ground looked as if it were covered with snow. But it wasn't snow, it was aspen seeds. The seeds had been carried by wind from outlying areas as well as from resident aspen groves, and had fluttered down onto the nitrogen-rich soil. The moss quickly grew in the burned areas, and the ash deposits retained important moisture. The aspen seedlings had no competition. Rocky Mountain aspen typically increase by sending out suckers of growth from "clones" or genetically identical groups of trees.

Roy Renkin remembered that new, genetically diverse aspen grew like grass that spring. By fall, Despain and Renkin measured as many as 1,000 aspen seedlings per square meter! Renkin says what amazed him was the astronomical number of seeds that landed in places they couldn't grow. Always the inquiring scientist, Renkin wondered if the prolific smoke of 1988 triggered a response by the aspen. Although the saplings have been winter-browsed by elk, all new aspen are surviving. Will they become new forests? Probably not. Competition and browsing will likely keep them the size of shrubs.

Rare Flora and Fauna

An abundance of Bicknell's geranium (*Geranium bicknellii*), a rare plant, was found post-fire. In Yellowstone, this flowering plant is biennial, which means it completes its life cycle in two years. In 1990, the plants bolted to a conspicuous height of two to three feet and then flowered. The next year they were nowhere to be found. The seed capsules lie in wait in the forest floor for the next big fire or disturbance, when their capsules will explode open again, flinging seeds from the last place the plant grew. The Bicknell's geranium seeds that germinated in 1989 may have lain in wait for two or three centuries!

Some birds also benefited from the fires, feasting on the proliferation of insects that took to fire-injured and dead trees. Also, their prey, such as mice and voles, were easier to spot from above. Two species of three-toed woodpeckers—three-toed woodpeckers and black-backed woodpeckers (*Picoides arcticus* and *P. tridactylus*)—that were rarely seen in the park before the fires were locally abundant during the first three years after the fires. Though not rare, the mountain bluebird and other cavity-nesters also benefited after the fires. Numbers of bald eagles as well as ospreys increased in the park after the fires, partly because raptors seek standing dead or burned trees to build their nests in. In recent years, many standing dead trees from 1988 have begun to fall, taking nesting sites with them, adversely affecting both eagle and osprey populations. The osprey population has declined more dramatically because their primary prey—young Yellowstone cutthroat trout—have been reduced by the non-native, predatory lake trout.

Declining Moose Populations

Biologist Dan Tyers studied moose in the Northern Range of the park, which is roughly composed of the Yellowstone and Lamar river drainages. Tyers determined that moose numbers in burned areas decreased dramatically, the only ungulate population that declined because of the 1988 fires. Moose had difficulty accessing subalpine fir, their winter food, through the deep snow in the burned areas. Instead, moose gnawed on the less palatable lodgepole pines

and almost three times as many willows as normal that first winter. They competed for forage with the elk that also browsed on the willows, which may have played a role in their waning numbers.

In the first five years after wolf reintroduction in 1995, thirteen moose succumbed to wolves. For whatever reason, moose populations have not recovered, and they are rarely seen in areas of the park where they once were common.

No Surprises: A New Fire Policy is Born

Not surprising, three congressional hearings were held after 1988 to review Yellowstone's fire policy. Experts determined that the size and intensity of the 1988 fires were underestimated by fire behavioralists because they could not properly assess the effects of prolonged drought and weather trends. The experts also concluded that Yellowstone fire managers needed more money, training, and experience. (Hindsight, after all, is always 20–20.)

Yellowstone is a land of many "firsts": It was the first national park in 1872; site of the first federal fire-fighting effort in 1886; one of the first parks to embrace a new natural fire program in 1972; and in 1988, Yellowstone was the first to test that "new" fire program.

After the fire season of 1988, Yellowstone was directed to suppress all fires—even though fire managers would argue that the summer's inferno wasn't a policy issue. In fact, after mid-July 1988, all fires were fought—it was just that drought and weather were in the driver's seat, and management took the back seat at best. As the American public and public lands fire managers questioned their fire policies, natural fire programs were suspended across the nation and even across the world.

Yellowstone's new fire policy was adopted in May 1992. The park would still allow natural fires to burn. Any fire—despite its cause—that threatened human life or a development would continue to be suppressed as it had before 1988. Park managers would have to consider the regional and national fire situation, how many fires were burning, and the availability of firefighters and equipment. Therefore, some lightning-caused blazes would be fought at their onset even if they appeared to be benign backcountry blazes. Overnight, an approved and monitored natural fire could be designated a wildfire, which would trigger full suppression. (And here's a paradox: If wildfires ignite in Yellowstone and there are already wildfires burning throughout the West, the park's fire personnel may be fighting fires elsewhere.)

Since 1988, Yellowstone's fire management staff has tripled in size. They have been assisted by computer models and real-time weather data that vastly improve their ability to predict fire behavior. In the field, National Park Service crews remove trees and debris from around buildings and developments as a preventative measure to minimize the risk of fire to park structures. Fire personnel monitor the forest's fuel moistures even before any flames begin, helping them to react appropriately if fire should come.

Because fire crosses human-made boundaries, fire managers are guided by one multi-agency document, the National Fire Plan, which manages fire from a big-picture perspective. Various fire experts also work with communities to ensure they are adequately protected from fire. ☒

Post–fire succession

What will Yellowstone National Park look like in another twenty years? Lodgepole pines will continue stretching skyward, growing taller and lankier, like telephone poles with Christmas trees on top. Eventually, the canopy of the forest will close as the tree branches begin to intertwine. As it becomes harder for sunlight to penetrate, the forest floor will grow darker, and the understory plants—forbs, wildflowers, and grasses—will begin to dwindle. At this stage,

other tree species propagated by birds, beasts, or wind will begin to thrive, plant species more tolerant of shade than the pioneering lodgepole.

Eventually, these forests will mature into mile after mile of tall lodgepole pines, resembling the forests that burned in 1988, and the forests that burned before that.

In the meantime, mature forests that didn't burn in 1988 will burn when the right conditions come together. This, of course, has already happened. The largest fire, the East Fire, burned in 2003 north of Yellowstone Lake and gives visitors an opportunity to track forest succession. The area between Fishing Bridge and the East Entrance still harbors the remnants of fire: patches of burnt ground interspersed with charred standing and fallen spruce and fir trees. But, hidden by lush understory plant growth are a scattering of tiny lodgepole seedlings. As time goes on, pine, spruce, and fir trees will emerge. The ancient cycle of death and rebirth begins again.

1988: Dunraven Pass was a landscape of burnt trees and ash in the wake of the 1988 fires in Yellowstone.

1992: Ecologist Don Despain and his staff analyze vegetation regrowth in the Washburn Mountain Range, five years after the 1988 burn.

1998: Ten years after the Wolf Lake Fire, lodgepole pine seedlings reemerge near Virginia Cascades Drive.

2004: Sixteen years after the North Fork Fire swept through, there is dramatic regrowth of lodgepole saplings in the Firehole River area.

Photos by Jeff Henry, Roche Jaune Pictures.

1992

1998

2004

Chapter Seven

A Burning Legacy

SEVEN

"It's that . . . sense of wildness that the parks were created to preserve and that unconquerable spirit which makes them great. And if we can learn not to hate or to fear that which is greater than ourselves then we, too, if only for a fleeting moment, can share in that glory."

—CAROL SHIVELY, "Trial by Fire," *Natural History* Magazine, 1989.

Yellowstone was born of fire, and, in 1988, Yellowstone was reborn by fire. Fire is integral to the park's landscapes and all its inhabitants—the pines, wildflowers, predators, and prey. Protect a forest's right to burn and you have protected the whole idea of Yellowstone. And what is that idea? Yellowstone National Park was preserved, in the words of the enabling legislation, "from injury or spoliation, of all timber, mineral deposits, natural curiosities, or wonders within said park. . . ."

Initially, it was the area's geological manifestations—the hot springs and geysers—that prompted the park's creation in 1872. These geological wonders were unaffected by the fires of 1988. Animals adapted to fire continued to graze or hunt while fires burned nearby. Curious visitors continued to experience the park during the summer of 1988 while firefighters fought fires throughout the park.

Although the fires were fought with unprecedented fervor in 1988, Mother Nature called the shots. We tried to control the fires, but, in the end, the fires controlled us. The processes of Yellowstone were protected by nature's own hand, ensuring that the park we all hold dear will not be the same. The changing adventure of Yellowstone is one of the reasons why many people return to the park year after year. Wiping out unpredictability, disallowing diversity, chance, and fire—even if it were possible—would be silencing nature and would destroy the wonder of the world's first national park.

Yellowstone has helped the world learn about fire, humility, and tolerance. Fire is necessary, despite the havoc and heartache it creates for us. Of course, there are places that people don't want fires to burn. Havens of human life: villages, towns, cities, and developments of any kind should be protected. Instead, to successfully live within any ecosystem that relies on fire, people have to shift focus from fire control to fire acceptance, to learn not only to live with fire, but to allow fire to function as it must.

Visiting Yellowstone National Park is magical, not predictable. Yellowstone is a place where people come to be renewed and refreshed. It is also a place where they come for the unexpected: Will they see a grizzly bear family today? Will a geyser that has lain dormant for many years erupt? Will a coyote find an elk calf? Will it snow in July?

There are no guarantees—only surprises to honor and enjoy, knowing that each visit to Yellowstone will spark new experiences and kindle new memories. Like a good story in the minds of the people, the 1988 fires will never die. ◪

Preceding pages: Norris Geyser Basin in 2006, where post-fire regeneration is well-established. Photo by Jeff Henry, Roche Jaune Pictures.

Facing page: Gibbon Falls plunges eighty-four feet over remnants of the Yellowstone Caldera. Photo by Jeff Henry, Roche Jaune Pictures.

Top: National Park Service photographer Jeff Henry walks across a burned area near Canyon. The Norris blowdown was created when a tornado swept through the area in 1984. These fallen logs contributed to intense burning in 1988, during the Wolf Lake Fire.

Photo courtesy of Larry Mayer, *The Billings Gazette*.

Above: Henry returned to the Norris blowdown on August 20, 1998, the ten-year anniversary of the infamous Black Saturday. Lodgepole pine seedlings and other vegetation are signs of the area's post-fire regeneration. Photo by Jeff Henry, Roche Jaune Pictures.

Right: Henry at the same spot in 2004. Note the height of the trees in comparison to the photographer.

Photo courtesy of Conrad Glenn Smith.